6th Triennial Symposium on the Acoustics of Poro-Elastic Materials (SAPEM 2020+1)

AA005986

Held online due to COVID-19

Lafayette, Indiana, USA
29 March - 2 April 2021

ISBN: 978-1-7138-3084-9

Printed from e-media with permission by:

Curran Associates, Inc.
57 Morehouse Lane
Red Hook, NY 12571

Some format issues inherent in the e-media version may also appear in this print version.

Printed with permission by Curran Associates, Inc. (2021)

Purdue University and Matelys – Research Lab
585 Purdue Mall
West Lafayette, Indiana, USA
47907-2088

https://engineering.purdue.edu/ME
https://www.matelys.com/

Additional copies of this publication are available from:

Curran Associates, Inc.
57 Morehouse Lane
Red Hook, NY 12571 USA
Phone: 845-758-0400
Fax: 845-758-2633
Email: curran@proceedings.com
Web: www.proceedings.com

Prosynx
Innovative Solutions through Sim-Test

MATELYS
RESEARCH LAB

3M

HBK
HOTTINGER BRÜEL & KJÆR

ALTAIR

SAPEM 2021
6TH TRIENNIAL
SYMPOSIUM ON THE ACOUSTICS OF PORO-ELASTIC MATERIALS

2_TUESDAY_MARCH_30 CHARACTERIZATION & MEASUREMENTS I	Tian Jian Lu	Yutong (Tony) Xue
Structural damping properties of natural porous material of jute felt...16	R. Kumar, A. R. Mohanty	Department of Mechanical Engineering, India
Characterization of the elastic parameters of porous materials as function of the initial load...20	F. Marchetti, F. Chevillotte	MATELYS-Research Lab, France
Equivalent dynamic model of multilayered structures with imperfect interfaces: application to a three-layer plate with sliding interfaces...21	N. Auquier, K. Ege, E. Gourdon	Univ. Lyon, France
Low frequency acoustic method to measure static thermal permeability...23	E. Di Giulio, M. Napolitano, F. Auriemma, R.A. Romano, R. Dragonetti	Department of Industrial Engineering, Italy
3M Sponsor session...25		
Matelys Sponsor session...29		

3_TUESDAY_MARCH_30 NEW MATERIALS	Michael Haberman	Marco Miniaci
Rapid additive manufacturing of optimised anisotropic metaporoussurfaces for broadband absorption...39	T. Cavalieri, J. Boulvert, V. Romero-Garcia, G. Gabard, M. Escouflaire, J. Regnard, and J.-P. Groby	LAUM, France
3D printed folded porous material for sub-wavelength absorption of sound...42	J. Boulvert, J. Costa-Baptista, E. R. Fotsing, T. Cavalieri, V. Romero-Garcia, G. Gabard, A. Ross, J. Mardjono, J.-P. Groby	LAUM, France
3D Printed PLA-hydogel fractal acoustic composite metamaterial...44	G. Comandini, F. Scarpa, M. Azarpeyvand, V. P. Ting	Bristol Composites Institute, UK
Natural sonic crystal absorber constituted of Aegagropilae fibernetwork...46	L. Barguet, V. Romero-Garcia, N. Jiménez, L.M. Garcia-Raffi, V. J. Sanchez-Morcillo, J.-P. Groby	LAUM, France
Sound insulation of clamped single panel and infinite double panel system by addition of poro-elastic resonators...48	Ke Li, N. Dauchez, B. Nennig	Université de technologie de Compiègne, France

ULTRALIGHT HYBRID-CORED SANDWICH CONSTRUCTIONS FOR SIMULTANEOUS LOAD BEARING, ENERGY ABSORPTION AND SOUND ABSORPTION

T. J. Lu

State Key Laboratory for Mechanics and Control of Mechanical Structures, Nanjing University of Aeronautics and Astronautics, Nanjing 210016, P.R. China

ABSTRACT Sandwich constructions with either two-dimensional (2D) prismatic or three-dimensional (3D) lattice truss cores, such as honeycombs, folded panels (corrugations) and pyramidal trusses, are ultralight multifunctional structures: in addition to sustain structural loading, they can simultaneously absorb impact energy, dissipate heat, and attenuate sound. These properties can be significantly improved further by inserting different materials (e.g., open/close-celled foams, miniature honeycombs, ceramics, sand, liquid, and so on) into the interstices of the lattices to construct a multitude of novel hybrid lattice-cored sandwiches, as demonstrated in this lecture. Particular focus is placed upon two different types of hybrid lattice-core (i.e., metallic/polymeric foam-filled lattice core and metallic honeycomb-corrugation core) for simultaneous load bearing, energy absorption and broadband low-frequency sound absorption. Physical mechanisms underlying the remarkable performance enhancements and multi-functionalities of hybrid-cored sandwich structures are systematically explored using a combined approach of theoretical modeling, numerical simulation and experimental testing.

Camille Perrot

(Université Gustave Eiffel, France)

What are the microstructural features having a significant effect at the upper scale in acoustic materials? Understanding the physics and guiding the manufacturing process

Abstract (not so short)

What is the microscopic basis of macroscopic transport in poro-elastic materials? How do macroscopic transport properties depend on pore-structure? These are two of the many questions that have dominated studies of sound waves through microscopically-disordered materials. In light of the manifest technological importance of determining the effective parameters of disordered heterogeneous materials (e.g. composite and porous media), an enormous body of literature has evolved based upon direct measurement, empirical relations, and approximate as well as rigorous theoretical methods [see Gibson and Ashby (1988); Adler (1992); Torquato (2001); Auriault et al. (2009) and references therein]. Performing direct measurements on each material sample, for possible multi-phase property values and all volume fractions, is prohibitive from a time and cost standpoint. Empirical relations are more useful for correlating data rather than predicting them. Since bulk properties are sensitive to the details of the microstructure, a broader approach is to calculate the properties from the microscopic structure (microstructure) of the disordered material; one can then relate changes in the microstructure quantitatively to changes in the macroscopic parameters. For general random media, the complexity of the microstructure prevents one from obtaining the effective properties of the system exactly. Some progress can be made with idealized structures. Idealized structures are useful since they: (i) enable one to test the merits of simple geometries and computer experiments; (ii) as successfully more geometrical information is incorporated, the effective property estimates become progressively more comparable with experiments; (iii) one of the microstructural model can typically provide a good estimate of the properties for a wide range of manufacturing conditions. In acoustics, the transport phenomena that are of interest are essentially heat conduction, inertial flow and viscous flow; together with elastic properties of a porous medium. Hence, the situation can be summarized as follows: field equations corresponding to four basic phenomena must be solved for three basic model geometries (cellular, granular, fibrous). The microscopic level corresponds to the knowledge of the value of the fields at every point. The macroscopic level is obtained by averaging the relevant microscopic fields over the virtual sample. It is this macroscopic level that is of interest for the engineers who wish to know the overall transport coefficients that relate the flux to the driving forces; and the effective elastic properties that enable the determination of the deformation under stress of the porous materials. One of the purpose of this communication is to provide a brief review of that popular micro-macro approach, that relate pore structure to transport coefficients (rigid assumption), so that their significance are clearly understood. Recent progress that we have made on the problem of determining the long wavelength effective acoustical properties of random heterogeneous media from knowledge of the morphology is reviewed and extended. This communication will primarily deal with non-woven and heterogeneous cellular structures.

Bhisham Sharma

(Wichita State University, USA)

Additive Manufacturing of Porous Noise Absorbers

Absorbing unwanted sound waves without excessively constraining the air flow path is a recurring challenge across various engineering fields. Cellular porous materials are the most commonly employed solution to this ubiquitous problem. The properties of traditional porous materials result from the stochastic architecture of their cellular solid phase—a product of conventional foaming techniques. While these materials provide reliable acoustical performance, a lack of control over their cellular structure limits their structural applications. The recent emergence of additive manufacturing has opened up the possibility of fabricating porous materials with cellular microstructural topologies that were previously infeasible. In this talk, I will showcase a few examples of additively manufactured, porous noise absorbers being investigated in our Mechanics and Dynamics Lab at Wichita State University including: polymeric and metallic foams with triply periodic minimal surfaces with unit cell dimensions ranging from 50 μm to 2 mm; 3D printed fibers to help improve acoustical performance of load-bearing lattices; and aerogel-based absorbers with controlled microstructures printed using a novel freeze-casting approach. These 'designer foams' present a pathway towards realizing application-tailored microstructures that can help improve sound absorption behavior while enabling a new paradigm of multifunctional engineering design.

Application of porous materials in the automotive industry when subject to multi-dimensional constraints

Gordon L. Ebbitt
Ebbitt Acoustical Consulting, LLC

The automotive industry purchases billions of dollars of noise and vibration (N&V) control materials and that amount is growing. Porous materials make up a significant portion of those noise and vibration materials and their use will likely grow even faster as the industry moves towards electric vehicles. Though this is a very exciting market for porous materials, there are a significant number of constraints that have to be considered before a promising new material goes into production. Of course, cost, weight, and N&V performance are the primary constraints, but there are others that are not as obvious. Those include engineering constraints such as manufacturability and physical properties. They also include commercial constraints such as the strategic goals of the auto manufacturer's purchasing department among many others. The process involved with introducing a new material to an automobile manufacturer will be discussed and the constraints that must be satisfied will be considered.

Wave propagation in hierarchical nanoporous materials

R. Venegas[1]
[1]University Austral of Chile, Institute of Acoustics, Valdivia, Chile.

Abstract

This talk will discuss wave propagation in hierarchical permeable materials characterised by sizes ranging from nanometres up to millimetres and beyond. Specifically, theories of wave propagation in i) nanoporous materials, ii) conventional porous materials with nanoporous coating, iii) double and triple porosity sorptive materials, and iv) porous composites with hierarchical nanoporous constituents will be presented. Particular attention will be devoted to the key role of several different types of diffusion processes and adsorption/desorption on the acoustical properties of hierarchical porous materials as well as how the introduction of pore hierarchies can turn nanoporous solids into efficient low frequency sound absorbers. The talk will conclude by discussing open problems in wave propagation in hierarchical nanoporous materials.

---

1 Email: rodolfo.venegas@uach.cl

Poro-elastic Materials and the Control of Low Frequency Sound

J. Stuart Bolton
Ray W. Herrick Laboratories
Purdue University
177 S. Russell St.
West Lafayette IN 47907-2099

In the introductory sections of active noise control and metamaterial articles, it is often said that "conventional", i.e., poro-elastic materials such as foams and fibrous media, do not work well at low frequencies. While that observation may be true for the simplest treatments, e.g., a single layer of a homogeneous, limp fibrous layer, there are many cases in which excellent weight and cost-effective acoustical treatments can be realized by using poro-elastic media. The first example involves the serendipitous discovery of a configuration that allows a 25 mm thick foam layer to provide effective absorption at 300 Hz, at a surface density substantially less than 1 kg/m^2. In the context of sound transmission, it will be shown that cells of edge-constrained fibrous media can yield astonishingly high transmission losses at low frequencies, say below 100 Hz, owing to a mechanism similar to that exploited in cellular membrane metamaterials. However, in both cases, a fair comparison with the performance of "conventional" barrier materials, i.e., simple impermeable mass layers, can only be drawn when the weight required to achieve the edge-constraint effect is accounted for.

Towards the design of soft periodic porous structures for acoustic applications

Sagar Deshmukh[1], Ankush Borkar[1], Alankar Alankar[1], Shankar Krishnan[1] and Sripriya Ramamoorthy[1*]

[1]Indian Institute of Technology Bombay, Mumbai, India

*Corresponding author email: ramamoor@iitb.ac.in

Abstract

Porous materials such as polyurethane foams, glass wool, jute, synthetic fibers are widely used in industrial noise control. It is well understood that the microstructure of the porous media largely affects its macroscopic response [1]. Due to the nature of the foaming process, it is difficult to control the microstructure of the foams beyond a certain limit and therefore it is difficult to obtain predictable macroscopic performance, especially in initial design phases. This fact has inspired the use of periodic foam or lattice structures for sound absorption [2]. Periodic foams are made of unit cell which is having periodic repetition in three dimensional domain as shown in Fig. 1.

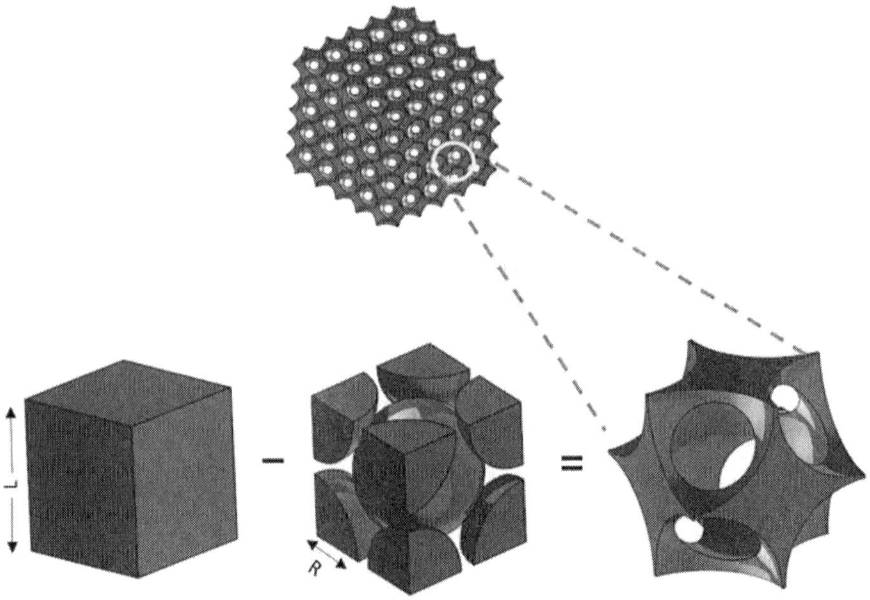

Fig. 1: Concept of periodic foam structure using body-centric (BCC) unit cell configuration

The periodic nature of these materials offers control over the microstructural properties of the porous material and thereby control over the macroscopic acoustic responses such as absorption coefficient and transmission loss. Recently, Deshmukh et al. [2] used the concept of periodic foam microstructure to obtain a predictable macroscopic response. Three different unit cell geometries viz. BCC, FCC, and A15 are used to create the periodic configuration and their macroscopic performance is compared based on the popular Johnson-Champoux-Allard (JCA) model. Later, these configurations are additively manufactured using polymer and metal as a base material in order to experimentally validate their macroscopic performance. The JCA model assumes the skeleton part of the porous material as elastically rigid (motionless) and therefore neglects the effect of structural modes on macroscopic acoustic response. However, porous materials with low elastic modulus, very small strand thickness, and/or high porosities, can have strong fluid-structure interaction, and therefore the rigid-skeleton assumption may lead to inaccuracies in predicting the responses of these soft materials.

In the present study, the use of periodic configurations to control the structural modal effects on sound absorption has been demonstrated based on Biot-uP formulation. Three unit cell configurations from Deshmukh et al. [2] are taken here for study. Their modulus values are computationally predicted using homogenisation approach. Predicted modulus values along with the microscopic JCA parameters of three configurations are used to carefully study the effects of structural modes on absorption coefficient. A parametric study on these three configurations has been performed by changing their porosity and the unit cell size. It is found that the effect of structural modes enhances with the increase in porosity as well as a decrease in unit cell size. However, it has been shown that the effect of structural modes on macroscopic acoustic response can be tuned (as per the application requirement) through careful selection of unit cell, porosity, and cell size. In order to prove the concept of *a priori* prediction of macroscopic poroelastic response, a periodic structure made of A15 unit cell configuration has been additively manufactured using a soft polymer material PIC 100 [3].

Two types of measurements (i.e. acoustical and structural) are carried out on the additively manufactured sample. In the first experiment, the absorption coefficient of the sample is measured using the two-microphone impedance tube method. Predicted absorption coefficient results based on the Biot-uP model are found in agreement with that of experimental results as shown in Fig. 16 in Deshmukh et al. [3]. However, due to the weak coupling between the air and fabricated sample as well as due to the higher value of loss factor, the effect of structural resonance is damped in the absorption coefficient graph. Therefore in the second experiment, vibrations of frame induced due to sound excitation are measured using accelerometer placed on front surface of the sample and results are shown in Fig.18 in Deshmukh et al. [3]. The measured frame displacement per unit incident pressure is found in agreement with that obtained using the Biot-uP model. The discrepancies between the Biot-uP model and the experimental results could be due to the manufacturing inaccuracies as well as difficulty in mimicking the structural boundary conditions during measurements [3]. However, *a priori* prediction of these acoustics and structural responses should be underlined once again, these macroscopic responses are predicted even before the fabrication of a sample and therefore the use of soft periodic porous structure can significantly reduce the design cycle time for material selection.

Acknowledgment:

This work was supported by IMPRINT-INDIA project no. 6367 and by the Department of Science and Technology grant no. AMT/620.

References

[1] C. Perrot, R. Panneton and X. Olny, "Periodic unit cell reconstruction of porous media: Application to open-cell aluminum foams," *Journal of Applied Physics,* no. 101, p. 113538, 2007.

[2] S. Deshmukh, H. Ronge and S. Ramamoorthy, "Design of periodic foam structures for acoustic applications: Concept, parametric study and experimental validation," *Materials & Design,* vol. 175, p. 107830, 2019.

[3] S. Deshmukh, A. Borkar, A. Alankar, S. Krishnan and S. Ramamoorthy, "A priori determination of the elastic and acoustic responses of periodic poroelastic materials," *Applied Acoustics,* vol. 169, p. 107455, 2020.

Experimental study of a meta-poro-elastic system with needle inclusions to improve the low-frequency absorption performance

S. Ahsani[1,2,3], Q. Zhang[3,4], B. C. Kim[4] C. Claeys[1,2], F. Scarpa[3,4], W. Desmet[1,2] E. Deckers[2,5]

[1] KU Leuven, Department of Mechanical Engineering, Division LMSD, Celestijnenlaan 300 - box 2420, Heverlee, Belgium
[2] DMMS Lab, Flanders Make, Heverlee, Belgium
[3] Dynamics and Control Research Group (DCRG), University of Bristol, BS8 1TR Bristol, UK
[4] Bristol Composites Institute (ACCIS), University of Bristol, BS8 1TR Bristol, UK
[5] KU Leuven, Campus Diepenbeek, Mechanical Engineering Technology, Wetenschapspark 27, 3590 Diepenbeek

This paper investigates experimentally the effect of embedding sub-wavelength scale inclusions in a foam layer on the sound absorption performance. It is shown that the absorption peak is due to the mass-spring effect, and is dependant on the inclusion mass and position.

1 Introduction

Poro-elastic materials suffer from lack of efficiency at low frequencies, despite their effective sound absorption behavior at higher frequencies. This motivated many researchers to focus on extending their application to lower frequencies, by means of meta-porous [1], and meta-poro-elastic systems [2], and benefiting from different effects, namely trapped mode effect [1], modified mode effect [3], and mass-spring effect [4]. The mass-spring effect exploits the frame vibration to increase viscous losses. So far, very limited research is dedicated to engineering meta-poro-elastic systems that benefit from the mass-spring effect. In this paper, a meta-poro-elastic design is proposed and validated experimentally, using the mass-spring effect as a mechanism to improve the sound absorption.

2 Meta-poro-elastic design

This work investigates meta-poro-elastic systems with small heavy inclusions in order to exploit the mass-spring effect. The mass-spring system has a specific resonance frequency and, beside modifying the mode(s) of the frame, it introduces an additional mode in the system.
The proposed design comprises a steel needle inclusion of diameter $r = 0.2$ mm inserted in a cylindrical foam sample of 40 mm thickness, and 40 mm diameter. The position of the needle in the foam layer is chosen such that the dependency of the mass-spring effect on the location of the inclusion, i.e. the stiffness of the spring, can be assessed. An schematic view of the two variations of the design are shown in Figure 1. Moreover, to evaluate the effect of the added mass on response of the system, design1-C is considered. In this configuration two needles are inserted at the same position as design1-B.

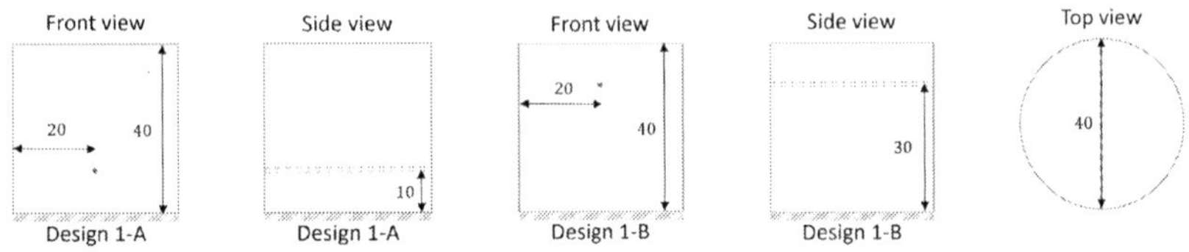

Figure 1: Schematic view of design 1-A(top), and design 1-B(bottom). The dimensions are given in millimeter.

3 Results and discussion

The absorption coefficient of the bare and meta-poro-elastic samples under normal incidence angle are shown in Figure 2. These values are measured using a Kundt tube that has a valid frequency range of 500 Hz to 4530 Hz, according to ISO 10534-2. Figure 2 shows that the needle inclusion causes the absorption peak of the bare foam to be split in two peaks. Furthermore, moving the inclusion further away from the rigid backing, shifts the first peak to lower frequencies, i.e. from 1100Hz for design1-A to 1000Hz for design1-B, due to due to an increased length of the foam which will be compressed, and hence a reduced stiffness. Moreover, a similar trend is noticed when the added mass is doubled. It should be noted that the efficiency of the mass-spring effect is highly dependent on the material properties of the foam since the additional mode induced by the resonance inducing inclusion puts the skeleton in out-of-phase motion with respect to the fluid phase and leads to an increase in viscous losses. Therefore, the presented results are considered as a proof-of-concept, and further improvement of the sound absorption is possible if the poro-elastic material is chosen wisely.

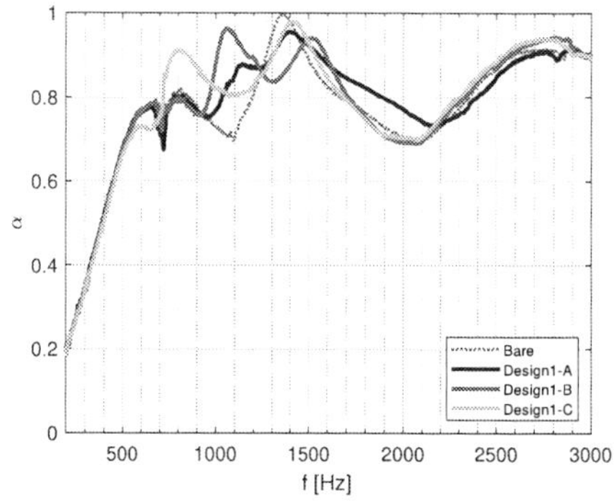

Figure 2: The absorption coefficients of the bare and meta-poro-elastic configurations measured using Kundt tube.

4 Conclusions

In this work, a meta-poro-elastic design with needle inclusions is presented to improve the sound absorption of the system at low frequencies, exploiting the mass-spring effect. It is demonstrated that using resonance inducing inclusions in an elastic-frame foam leads to an additional absorption peak. Moreover, it is experimentally validated that this absorption peak is tunable by tweaking the position and the mass of the inclusion.

Acknowledgements

The European Commission is gratefully acknowledged for their support of the VIPER research project (GA 675441). The Research Fund KU Leuven is gratefully acknowledged for its support. The research of E. Deckers is funded by a grant from the Research Foundation – Flanders (FWO).

References

[1] C. Lagarrigue, J. P. Groby, V. Tournat, O. Dazel, O. Umnova, Absorption of sound by porous layers with embedded periodic arrays of resonant inclusions, J. Acoust. Soc. Am. 134 (6) (2013) 4670–4680.

[2] T. Weisser, J.-P. Groby, O. Dazel, F. Gaultier, E. Deckers, S. Futatsugi, L. Monteiro, Acoustic behavior of a rigidly backed poroelastic layer with periodic resonant inclusions by a multiple scattering approach, J. Acoust. Soc. Am. 139 (2) (2016) 617–629.

[3] J.-P. Groby, O. Dazel, A. Duclos, L. Boeckx, L. Kelders, Enhancing the absorption coefficient of a backed rigid frame porous layer by embedding circular periodic inclusions, J. Acoust. Soc. Am. 130 (6) (2011) 3771–3780.

[4] S. Ahsani, E. Deckers, T. G. Zielinski, L. Jankowski, C. Claeys, W. Desmet, Absorption enhancement in poro-elastic materials by mass inclusion, exploiting the mass-spring effect, in: Proc. of

SMART'2019 9th ECCOMAS Thematic Conference on Smart Structures and Materials, 2019, pp. 1076–1084.

Towards elastic wave manipulation in Kelvin Cell packings with controlled anisotropy

Mathieu Gaborit[1,2], Romain Rumpler[1,2], Huina Mao[1,2], Peter Göransson[1,2]

[1]The Marcus Wallenberg Laboratory for Sound and Vibration Research (MWL), Department of Engineering Mechanics, KTH Royal Institute of Technology, SE-100 44, Stockholm, Sweden
[2]The Centre for ECO2 Vehicle Design, SE-100 44 Stockholm, Sweden

The present contribution proposes to reflect on the use of packings of distorted Kelvin Cell to create controlled anisotropic media for elastic waves. The work is still in an exploratory stage and we will present the early investigations and results. The key aspects of the proposed research is to understand how one can modify the cell geometry to alter the slowness surfaces and propagation characteristics for the resulting media. On the long run, the idea is to incite interesting effects such as wave trapping, steering and controlled compression-to-shear conversion that can find application in a number of engineering domains.

1 Context

With new manufacturing capabilities come new strategies to design and construct novel complex materials. In the context of porous media, such materials are often periodic and the recent contributions have focused primarily on media mimicking a controlled pore network. It was shown that the proposed media can then be 3D printed and that measurements and simulations for airborne propagation agree well [1].
The design of complex structures obviously goes beyond fluid acoustics applications and particularly spans over the field of elastic waves. An option to construct controlled-anisotropic lightweight media is to use arrays of distorted Kelvin cells [2, 3]. In these media, the cell to be repeated corresponds for instance to a Kelvin cell such as the one presented on Fig. 1a with some of the square faces rotated (see Fig. 1b). This kind of modifications of the cell allow to generate anisotropic media and particularly, work is ongoing to control this anisotropy. One of the key aspects today is to understand how specific types of alteration lead to different patterns in the Hooke's tensor and harness these effects. Of course, other modifications of the cell can be proposed, leading to different effects which is the scope of another contribution by the authors.

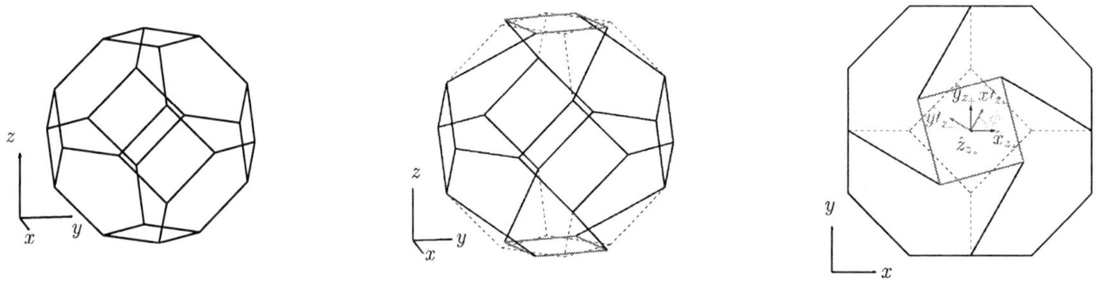

(a) Non distorted (b) Top and bottom faces rotated (c) Top view of Fig. 1b

Figure 1: The original Kelvin Cell undistorted (Fig. 1a) and distorted by rotating the top and bottom face around the z axis by 60 degrees (Figs. 1b and 1c). Reproduced from Ref. [3].

2 Proposed work

In this presentation, we will discuss the use of this new class of anisotropic media to create specific effects for elastic waves. The key aspect of the proposed research is to understand how one can modify the cell geometry to alter the slowness surfaces and propagation characteristics for the resulting media.

Being able to design groups of cells that together favour compression-to-shear conversion for instance, paves the way to a number of systems capable of trapping waves and dissipating the incoming energy within a short distance. Controlling entirely the anisotropy of a given medium also allows to target specific modes and answer precise design requirements.
The path towards such applications is arduous and this presentation aims at giving an overview of the strategy we are implementing. This includes for instance the construction of surrogate models to mitigate the computational costs of performing the numerical characterisation of the media [4] or the approach taken to identify promising alterations for a given target effect. The work is still in an exploratory stage and we intend to present the early investigations and results, hoping to foster discussions and comments.

References

[1] Tomasz G. Zielinski, Kamil C. Opiela, Piotr Pawlowski, Nicolas Dauchez, Thomas Boutin, John Kennedy, Daniel Trimble, Henry Rice, Bart Van Damme, Gwenael Hannema, Rafal Wrobel, Seok Kim, Shahrzad Ghaffari Mosanenzadeh, Nicholas X. Fang, Jieun Yang, Baltazar Briere de la Hosseraye, Maarten C.J. Hornikx, Edouard Salze, Marie-Annick Galland, Rene Boonen, Augusto Carvalho de Sousa, Elke Deckers, Mathieu Gaborit, and Jean-Philippe Groby. Reproducibility of sound-absorbing periodic porous materials using additive manufacturing technologies: Round robin study. *Additive Manufacturing*:101564, September 2020. DOI: 10.1016/j.addma.2020.101564.

[2] H.J. Rice, J. Kennedy, P. Göransson, L. Dowling, and D. Trimble. Design of a Kelvin cell acoustic metamaterial. *Journal of Sound and Vibration*, 472:115167, April 2020. DOI: 10.1016/j.jsv.2019.115167.

[3] Huina Mao, Romain Rumpler, Mathieu Gaborit, Peter Göransson, John Kennedy, Daragh O'Connor, Daniel Trimble, and Henry Rice. Twist, tilt and stretch: From isometric Kelvin cells to anisotropic cellular materials. *Materials & Design*, 193:108855, August 2020. DOI: 10.1016/j.matdes.2020.108855.

[4] Huina Mao, Romain Rumpler, and Peter Göransson. An inverse method for characterisation of the static elastic Hooke's tensors of solid frame of anisotropic open-cell materials. *International Journal of Engineering Science*, 147:103198, February 2020. DOI: 10.1016/j.ijengsci.2019.103198.

Comprehensive Evaluation of Microstructure Properties of Polyurethane Foams

T. Fitzgibbons, D. Monaenkova, C. Broomall, M. Chyasnavichyus, T. Fowler, L. Wright, D. Honkomp, K. Kiszka, J. Frketic, M. Allen, M. Thota

Dow Chemical Company, Corporate R&D

1 Abstract

It is well known that sound absorption coefficient of porous material is strongly dependent on the micro-structure of porous material. In this work, we evaluate a comprehensive list of micro-structure properties such as cell size & distribution, open porosity, closed cell fraction, closed face fraction and strut geometry of polyurethane foam. X-Ray tomography, mercury porosimeter, scanning electron microscopy (SEM) and other techniques available through the Dow MobilityScience™ platform are utilized for this evaluation. A variety of commercially available chemicals are formulated and foams are made in lab at different densities; the micro structure properties and acoustic absorption performance of such foams is discussed in this work.

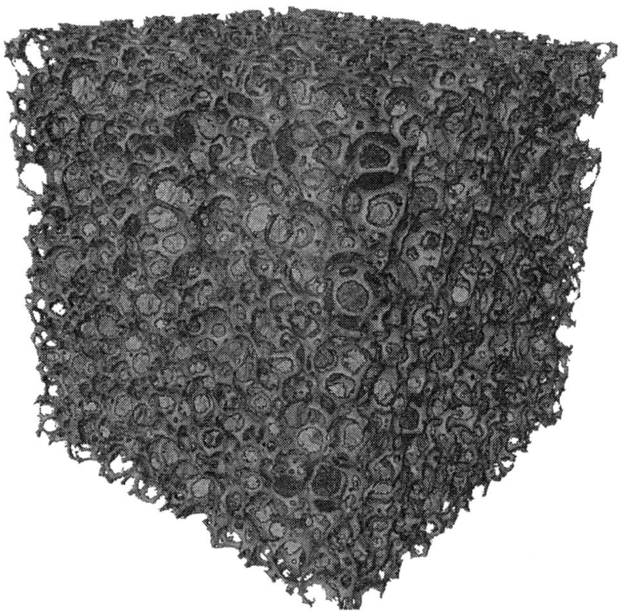

Can a perfectly matched layer be mimicked with a porous medium?

D. Keith Wilson, Michael B. Muhlestein, and Carl R. Hart

U.S. Army Engineer Research and Development Center
72 Lyme Rd.
Hanover NH 03755-1290

Corresponding author: Keith Wilson, d.keith.wilson@erdc.dren.mil

Abstract:

An earlier publication [D. K. Wilson and M. B. Muhlestein, *Proceedings of Meetings on Acoustics* 39, 045004 (2019), https://doi.org/10.1121/2.0001208] discussed the possibility of designing the properties of a porous material to mimic a perfectly matched layer (PML). (A PML is a numerical construct, providing essentially zero reflection and rapid attenuation in space.) The formulation for the acoustical properties of the medium was based on the relaxation model, from which it was shown that setting the ratio of the vorticity and entropy shape factors to $\sqrt{N_{pr}}/(\gamma-1)$, where N_{pr} is the Prandtl number and γ the ratio of specific heats, the impedance becomes real-valued while the medium remains attenuative. However, this pore-shape factor ratio is approximately 2.1 in air, whereas it appears that for typical pore geometries it must be bounded between 0 and 1. Therefore the PML-like condition does not appear to be physically realizable. In this presentation, we summarize the previous derivation and discuss whether extensions to closed-pore geometries are capable of producing the desired pore-shape factor ratios larger than one.

Structural damping properties of natural porous material of jute felt.

R. Kumar[1] and A. R. Mohanty[1]

[1]Department of Mechanical Engineering, Indian Institute of Technology,

Kharagpur, India

Abstract

Most sound-absorbing and damping materials are synthetic which are harmful to health of living beings and environment [1]. Worldwide with researcher's growing interest in utilizing porous material as the sound-absorbing and structural damping material, the focus is on developing natural, biodegradable, and recyclable porous material. Jute is a natural lignin cellulose-based material with great potential as a sound-absorbing material, as shown in studies [2-6]. Fatima et al. [2] showed that the low-density jute composite was a better sound-absorbing material than the high-density jute composite material. Bansod et al. [3] measured the acoustical properties of jute felt and jute fiber using experimental, analytical, and numerical methods. They extended this work to use jute with micro-perforated panel (MPP) and air gap to boost the sound absorption in the low-frequency region [4]. The work included the measurement of non-acoustic properties to be used in the Transfer Matrix Method (TMM) to characterize the sound absorption of jute with MPP and air gap considering different combinations. Later in [5], jute's acoustic properties are measured using the impedance tube transfer method. The measured property is then used to inverse characterize the porous material model using Particle Swarm Optimization (PSO) to obtain the non-acoustic parameter, i.e., airflow resistivity, porosity, thermal and viscous characteristics length of the jute felt. Mohanty et al. [6] used the jute as sound-absorbing material in a domestic clothes dryer's casing panel to reduce the noise emitting from the motor, belt, and casing of the dryer by around 6 dB. The noise produced in the dryer is the combination of the vibrations of the running motor, motor and belt, casing panel. The application of jute such as this and similar cases like metro rail, rail wagon, and aircraft where the vibrating structure is causing noise, utilizes the porous material damping property of the material. Hence it is necessary to characterize the structural damping property of the jute for its multipurpose use. Therefore, a need to review the work of structural damping material of porous material is required to choose the proper damping model to characterize its damping property. Wahl and Bolton [7] considered the thin Euler Beam panel in space and time domain under line force. The beam's governing equation is double Fourier transformed in space and time domain to obtain the wavenumber and frequency domain's governing equation. Then Discrete Fourier Transform (DFT) is used to find the panel's velocity in wavenumber frequency domain response considering the boundary conditions. Inverse Discrete Fourier transform (IDFT) is then used in the space domain to get the panel's spatial response. The work also presents the precautions to utilize the Fast Fourier transform (FFT) method to capture the spatial resolution correctly. They extended this work in [8] to evaluate the panel's temporal response by again Inverse Discrete Fourier transform method (IDFT). Xue et al. in [9-10] further used this work to assess the structural damping in terms of power dissipation and structural loss factor for the unconstrained and constrained infinite panel loaded with limp, bonded, and unbonded poroelastic material. The authors used this to design the microstructure of porous material for optimum damping to be obtained at a particular frequency. Further, Xue et al. in [11] did a parametric study based on the macroscopic properties (thickness, airflow resistivity, and bulk density) of glass fiber and polymeric fiber based on the damping model introduced in [9]. From the literature review, it is clear that jute is an excellent sound-absorbing material. If the jute shows potential as the structural damping material, it could save weight and cost. Therefore, this study's purpose is to use the natural material as structural damping material that is already proven sound-absorbing, cheap, and biodegradable material. It is a naturally porous material that can reduce the adverse effect by utilizing synthetic material environments. It is to be noted here that the methodology adopted for this study is from [9, 10] for the constrained and unconstrained panel for the damping by the jute felt as a limp porous layer. The present work is grouped in two. In group 1. (a), we study the jute's effect with an unconstrained panel. This includes evaluating the panel vibration response with or without increasing the thickness and bulk density simultaneously of the jute felt on the panel's vibration response, power dissipation within the porous layer, and structural damping loss factor. (b) The effect of the panel material (Aluminum and Stainless steel, which is generally used in the metro and rail wagons[11]) and thickness is studied as expressed in normalized power dissipation within a porous layer. c) Effect of porosity on normalized power dissipation within the jute felt. In the second Group 2, the panel's velocity and structural damping loss factors are evaluated for the jute felt. The Aluminium and Stainless steel panels density is 2100 and 7955 kg/m^3, the modulus of elasticity is 70 GPa and 203 GPa, the Poisson ratio is 0.33 and 0.27, and the loss coefficient is 0.003 and 0.00125. It is here to be noted that jute's macroscopic properties are taken from the previous work [5]. The jute felt of thickness 4, 8, and 10 mm with a bulk density of 76.13, 116.4, and 148.3 kg/m^3, airflow resistivity of 32751, 20803, and 28629 Ns/m^4, the tortuosity of 1.05 for each, the thermal characteristics length of 50, 67.45, and 50 μm and viscous characteristics length 129.85, 130, and 130 μm.

This study evaluates the damping effect for an unconstrained and constrained panel subjected to line force applied in Group 1 and Group 2, respectively. In Group 1, the unconstrained panel's velocity response vs. space is evaluated with and without the jute felt to compare its effect. The result shows that jute's use in the panel reduces the velocity response. As the thickness and bulk density of the jute felt increases, the panel's velocity response is reduced considerably faster before the critical frequency. The normalized power dissipation in the jute layer increases with the increase of thickness of the Jute layer. The difference between the first peak to the first trough of normalized power dissipation reduces, i.e., the normalized power dissipation increases at the trough, leading to more uniform power dissipation within the layer. The use of the Jute layer shows an appreciable damping loss factor. Increasing bulk density and thickness increases the damping loss factor at lower frequencies. The result also shows the panel thickness's effect; as the panel thickness is reduced, the critical frequency increases, but the difference between the peak and trough of the normalized power dissipation is decreasing, i.e., more uniform. The result also shows that the porosity variation has no significant effect on normalized power dissipation in the jute material before its critical frequency. Using the Stainless steel and Aluminium panel, the normalized power dissipation and damping loss factor in a porous layer are compared with the panel's same thickness. In the second group, the panel's velocity and structural damping loss factors are evaluated for the jute felt. The results are promising to utilize the jute felt as a damping material. Thus it paves the purpose of this study to use jute as a natural material for structural damping. Therefore Jute Felt could be utilized as sound-absorbing as well as structural damping material without causing harm to the environment.

References

[1] Asdrubali, Francesco; D'Alessandro, Francesco; Schiavoni, Samuele (2015). *A review of unconventional sustainable building insulation materials. Sustainable Materials and Technologies, 4(), 1–17.* doi:10.1016/j.susmat.2015.05.002

[2] S. Fatima; A.R. Mohanty (2011). *Acoustical and fire-retardant properties of jute composite materials. , 72(2-3), 108–114.* doi:10.1016/j.apacoust.2010.10.005

[3] Bansod, Pritesh Vishwasrao; Mittal, Tushar; Mohanty, Amiya Ranjan (2016). *Study on the Acoustical Properties of Natural Jute Material by Theoretical and Experimental Methods for Building Acoustics Applications. Acoustics Australia, 44(3), 457–472.* doi:10.1007/s40857-016-0073-4

[4] Bansod, Pritesh V; Teja, T Sai; Mohanty, Amiya R (2017). *Improvement of the sound absorption performance of Jute felt-based sound absorbers using micro-perforated panels. Journal of Low Frequency Noise, Vibration and Active Control, 36(4), 376–398.* doi:10.1177/1461348417744307

[5] Bansod, P.V.; Mohanty, A.R. (2016). *Inverse acoustical characterization of natural jute sound absorbing material by the particle swarm optimization method. Applied Acoustics, 112(), 41–52.* doi:10.1016/j.apacoust.2016.05.011

[6] Mohanty, A.R. & Fatima, Shahab. (2015). Noise Control Using Green Materials. Sound and Vibration Magazine. 49.

[7] T.J. Wahl and J.S. Bolton, "The Use of the Discrete Fourier Transform to Calculate the Spatial Response of Damped Panels," Proceedings of the Acoustics '90, 1990 Spring Conference of the Institute of Acoustics (U.K.), published in Proceedings of the Institute of Acoustics, Vol. 12 (1), pp. 945-952, 1990, https://docs.lib.purdue.edu/herrick/123.

[8] T. J. Wahl and J. S. Bolton, *The Use of the Discrete Fourier Transform to Calculate the Spatial and Temporal Response of Line-Driven, Layer-wise Homogeneous Acoustically Loaded Panels,* Journal of the Acoustical Society of America, Vol. 92, pp. 1473-1488, 1992 .

[9] Xue, Yutong; Bolton, J. Stuart (2018). *Microstructure design of lightweight fibrous material acting as a layered damper for a vibrating stiff panel. The Journal of the Acoustical Society of America, 143(6), 3254–3265.* doi:10.1121/1.5038255

[10]Xue, Yutong; Bolton, J. Stuart; Herdtle, Thomas; Lee, Seungkyu; Gerdes, Ronald W. (2019). *Structural damping by lightweight poro-elastic media. Journal of Sound and Vibration, (), 114866–.* doi:10.1016/j.jsv.2019.114866

[11] Yutong Xue, J. Stuart Bolton, Thomas Herdtle, Seungkyu Lee and Ronald W. Gerdes, "A Comparison between Glass Fibers and Polymeric Fibers when Serving as a Structural Damping

Medium for Fuselage-like Structures," *Proceedings of InterNoise 2018*, Chicago, Paper 1478, August 2018, https://docs.lib.purdue.edu/herrick/179.

[12] Government of India ministry of urban development, a report of the committee on "*standardization and indigenization of metro railways, systems and sub-systems.* November 2013

Characterization of the elastic parameters of porous materials as function of the initial load

F. Marchetti[1], F. Chevillotte[1]

[1] Matelys, 7 rue des Maraîchers, Bât B, 69120, Vaulx-en-Velin, France

The characterization of the elastic parameters is essential to feed the models and guarantee vibroacoustic performances. It is therefore necessary to have robust, comparable and reliable methods. Several characterization techniques exist and may lead to large differences between the results. Some inter-laboratory tests [1] have already been carried out in order to compare the methods. The main deviations are due to:

- the method used (quasi-static compression, resonant mass, dynamic torsion, Oberst beam). The material may not be excited the same way between the methods (traction/compression, bending, shearing, torsion). The results are strongly different between all methods especially if the material is anisotropic.
- the frequency range used. Since the mechanical properties depend on frequency, the characterization methods should be operated in the same frequency range. The viscous effect of the saturating fluid can also affect the results at high frequencies (over 100 Hz). This problem can be avoided measuring in vacuum.
- the initial loads imposed on the material. The load is either imposed on the stress or the strain depending on the method.
- the difference between apparent and intrinsic Young's moduli. Contrarily to the intrinsic Young's modulus, the apparent one depends on the shape factor of the sample and the Poisson's ratio of the material [2].

Recently, a more constrained Round Robin test was conducted with four laboratories in order to draw some guidelines [3]. On one hand, the goal of the characterization is to be able to compare the values between the laboratories. On the other hand, the load cases have to be close to the final application which can be significantly different between one material to another. The employed methods were based on uni-axial compression (quasi-static or spring mass methods). The intrinsic Young's modulus and the Poisson were systematically computed. The frequency range have been limited in the range 10-100 Hz while allowing an extrapolation. It has been pointed out that different initial loads or deformations can lead to large differences. They conclude that the compression rate and the initial load have to be estimated and clearly mentioned. An improvement could be to characterize the elastic properties for different compression rates.
The objective of this work is to investigate some methods to estimate the elastic parameters depending on the initial load (imposed stress or imposed deformation). Two methods are presented based on the quasi-static method and compared. The intrinsic properties are identified as the function of the initial stress or deformation. Then the results are compared and discussed.

References

[1] P. Bonfiglio et al., *How reproducible are methods to measure the dynamic viscoelastic properties of poroelastic media?*, Journal of Sound and Vibration **428**, pp. 26-43, 2018.

[2] C. Langlois, R. Panneton and N. Atalla, *Polynomial relations for quasi-static mechanical characterization of isotropic poroelastic materials*, J. Acoust. Soc. Am. **110**, pp. 3032-3040, 2001.

[3] F. Chevillotte et al., *Inter-laboratory characterization of Biot parameters of poro-elastic materials for automotive applications*, ISNVH, Graz, Austria, 2020.

Equivalent dynamic model of multilayered structures with imperfect interfaces: application to a three-layer plate with sliding interfaces

N. Auquier[1,2,3], K. Ege[2,3], E. Gourdon[1,3]

[1] Univ Lyon, ENTPE, LTDS UMR CNRS 5513, 3 rue Maurice Audin, 69518 Vaulx-en-Velin CEDEX, France - nicolas.auquier@entpe.fr, emmanuel.gourdon@entpe.fr

[2] Univ Lyon, INSA-Lyon, LVA EA677, F-69621 Villeurbanne, France - kerem.ege@insa-lyon.fr

[3] LabEx CeLyA - Centre Lyonnais d'Acoustique, 36 avenue Guy de Collongue, 69134 Ecully, France

This works aims to build an equivalent model of multilayered structures with imperfect interfaces for vibro-acoustic characterization. To do so, new boundary conditions containing the imperfections are implemented. The dispersion relation is derived from the equivalent model and the bending wavenumber is obtained. In this work we applied the methodology to a sandwich-structured composite with sliding interfaces. We computed the flexural rigidity and the damping of the flexural waves to see their effects.

1 Context and objective

This work aims to develop an equivalent model able to characterize and optimize the behavior of multilayered structures with imperfect interfaces and particularly a sandwich-structured composite, mainly used in the industry because of its high ratio stiffness over mass. Equivalent plate models are increasingly developed in the scientific community these last years [1, 2, 3, 4], because of their several advantages. For instance: a minimum calculation time allows the testing of many configurations efficiently *i.e.* without using a finite element model mainly 3D with a significant mesh; an easy-to-use method; physical meaning brought by the physical model. The goal of this work is to extend the latest improvement of the Guyader's model [5] done by Marchetti [6] for composite plates, by taking into account imperfect interfaces, and more specifically as a first step sliding between the layers.

2 Methodology

Previously, in Guyader's and Marchetti's work, continuity conditions were formulated for transversal shear stresses and the 3D displacement field. These conditions yield the expression of the cinematic variables of a layer in terms of the previous one, thanks to a coupling matrix. Finally, the global behavior of the composite is described by using only 5 cinematic variables of the reference layer. Thus, to take into account imperfect interfaces with non-canonical coupling conditions, we modified the continuity conditions of the displacements into conditions introducing imperfect interfaces, such as sliding or delaminations alike Massàbo's work [7]. An interface is modeled as a zero thickness layer, which contains local perturbations of the layer n written as jumps of displacement symbolized by

$$\hat{u}^n = u^n - u^{n-1}. \tag{1}$$

Therefore, a constitutive equation of the interfaces is added to compute the sliding displacement $\hat{u}^n$ with respect to the shear stresses applied at the interface

$$(\hat{\sigma}^n) = [K^n].(\hat{u}^n) + (t^n), \tag{2}$$

where the interface stresses $\hat{\sigma}^n$ is written as a piecewise linear function in order to be able to model many different types of interfaces, such as elastical or dissipative. K^n is the stiffness of the interface n and t^n the dissipation. The displacements, the interface stress and interface stiffness are illustrated in Fig. (1a). Once the displacement field of the reference layer is computed, the principle of least action is applied and leads to a system of differential equations, equivalent to a mass matrix and a stiffness matrix. Solving the determinant of the equivalent matrix yields the different wavenumbers according to the different types of movements. They are sorted to retrieve only the solutions corresponding to the bending wavenumber. From now, several equivalent parameters can be evaluated, such as the equivalent flexural rigidity, or the equivalent damping (loss factor) for further analysis.

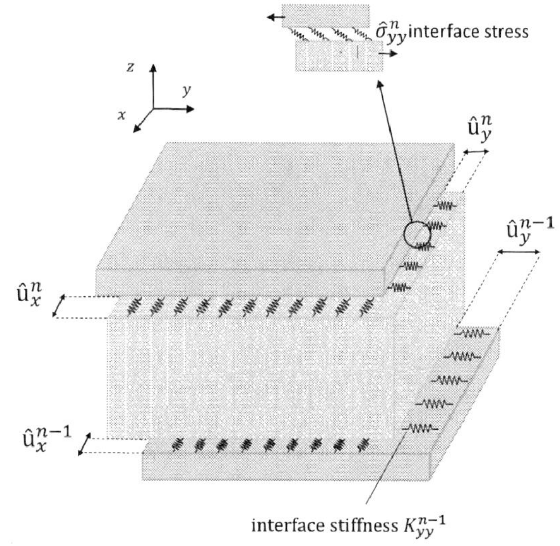

(a)

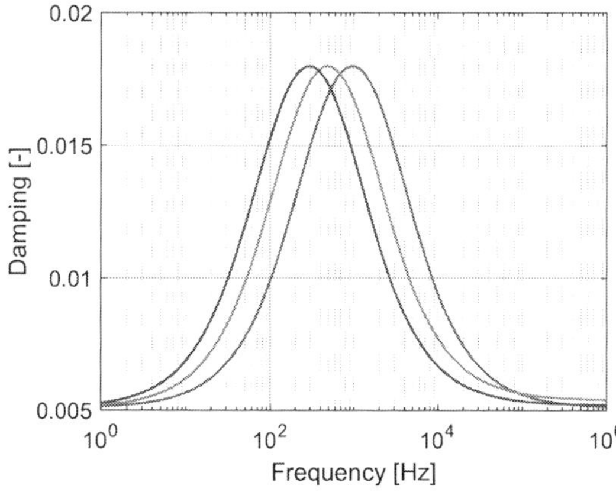

(b)

Figure 1: (a) Typical three-layered composite with a soft and thick core and illustration of the sliding with the generated interface stiffness, (b) Equivalent loss factor with perfect interfaces (–) and with imperfect interfaces: one imperfect interface ( ), two imperfect interfaces (–).

3 Results

We show in this work the impact of sliding interfaces on the equivalent flexural rigidity and damping of a typical three-layered composite with a thick and soft core (Fig. 1b). The impact is mainly a shift of the frequency of optimum damping towards the low frequencies, as well as a shift of the transition zone for the flexural rigidity, which is governed by the shearing of the core. The behavior of the composite stay unchanged in the low frequencies. For the high frequencies, the contribution of the sliding is not important enough to be physically valuable. Other effects and the asymptotic behaviors will be investigated and presented during the communication.

References

[1] E. Carrera, *An assessment of mixed and classical theories on global and local response of multi-layered orthotropic plates*, Composite Structures 50 (2), 183–198, (2000).

[2] S. Ghinet, N. Atalla. *Modeling thick composite laminate and sandwich structures with linear viscoelastic damping.* Computers & Structures 89, 547-1561, (2011).

[3] O. Dazel, J.-P. Groby, B. Brouard, C. Potel. *A stable method to model the acoustic response of multilayered structures.* Journal of Applied Physics 113, 083506, (2013).

[4] K. Viverge, C. Boutin, F. Sallet. *Model of highly contrasted plates versus experiments on laminated glass.* International Journal of Solids and Structures 102-103, 238-258, (2016).

[5] J.L. Guyader, C. Lesueur. *Acoustic transmission through orthotropic multilayered plates, part I: Plate vibration modes.* Journal of Sound and Vibration 58, 51-68, (1978).

[6] F. Marchetti, K. Ege, Q. Leclère, N.B. Roozen. *On the structural dynamics of laminated composite plates and sandwich structures; a new perspective on damping identification.* Journal of Sound and Vibrations 474, 115256, (2020).

[7] R. Massàbo, F. Campi. *Assessment and correction of theories for multilayered plates with imperfect interfaces.* Meccanica 50, 1045–1071, (2015).

Low frequency acoustic method to measure static thermal permeability

E. Di Giulio[1], M. Napolitano[1], F. Auriemma[2], R.A. Romano[1], R. Dragonetti[1].
[1]Department of Industrial Engineering, University of Naples Federico II, 80125 Naples, Italy
[2]Department of Mechanical and Industrial Engineering, Tallinn University of Technology - TalTech, 19086 Tallinn, Estonia

1 Introduction

Acoustic behaviour of air-saturated rigid-frame porous material is described by equivalent fluid model through complex density $\tilde{\rho}$ and bulk modulus $\tilde{K}$. Their experimental characterization can be made in impedance tube through 3-microphone or 4-microphone techniques. From an analytical point of view, semi-phenomenological models are commonly used to describe the sound propagation inside solid matrix filled by a fluid medium in terms of equivalent fluid [1-3]. The accuracy of each model is strictly linked to the number of input parameters which describe thermal and viscous dissipation phenomena inside the core. These so-called transport parameters can be evaluated through three different methods: an indirect one, where parameters are obtained from the analytical expression of complex density and bulk modulus, an inverse one, based on optimization procedure to match acoustic measurements and semi-phenomenological model, and a direct one, based on their definitions. In particular, to evaluate experimentally static viscous permeability k_0 (equivalent to static air flow resistivity σ) and static thermal permeability k_0' by using their definitions, low frequency measures are requested. In literature, an experimental alternative method to calculate static viscous permeability k_0 has been proposed [4]. In this paper, the aim is to extend this methodology for the calculation of static thermal permeability k_0'.

2 Theoretical formulation

The static viscous permeability k_0 can be derived as the low frequency limit of dynamic permeability $\tilde{k}(\omega)$. Combining Darcy's law with solution of linearized equation of motion, it can be written

$$k_0 = \lim_{\omega\to 0} \tilde{k}(\omega) = \lim_{\omega\to 0} \frac{\delta_v}{2i}\frac{\rho_0}{\tilde{\rho}_{eq}} \tag{1}$$

where $\delta_v = \sqrt{2\mu/\omega\rho_0}$ is the viscous penetration depth, ρ_0 the fluid density and $\tilde{\rho}_{eq}$ the equivalent complex density of porous medium. A thermal analogue of Darcy's law is defined by Lafarge, where a static thermal permeability $k_0' = \lim_{\omega\to 0} \tilde{k}'(\omega)$ is introduced to improve the description of low frequency behaviour of dynamic bulk modulus $\tilde{K}$. Operating the limit for the low frequency, static thermal permeability results from solution of linearized equation of energy and Thermal Darcy's law

$$k_0' = \lim_{\omega\to 0} \tilde{k}'(\omega) = \lim_{\omega\to 0} \frac{\gamma}{\gamma-1}\frac{\delta_\kappa}{2i}\frac{p_m}{\tilde{K}_{eq}} \tag{2}$$

with $\delta_\kappa = \sqrt{2k/\omega\rho_0 c_p}$ the thermal penetration depth, p_m the atmospheric pressure and $\tilde{K}_{eq} = \varphi\tilde{K}$ is the equivalent complex bulk modulus.

3 Static thermal permeability measurement

Starting from the test configuration suggested by Dragonetti et al. to measure the air-flow resistivity through "indirect acoustic method" (Figure 1.a), the aim of this work is to derive an acoustic methodology for the estimation of static thermal permeability k_0'.Hayden and Swift [5] proposed an experimental methodology to evaluate thermal Rott's function f_κ, which is linked to complex bulk modulus $\tilde{K} = \gamma p_m/(1+(\gamma-1)f_\kappa)$. Their apparatus was made of an oscillating piston which produced changes in the volume of a cylindrical chamber. This method required an upscale of the geometry of the porous material, the estimation of pressure beyond the tested sample and the volume perturbation with and without the material, respectively $(V_1, p_1)_{full}$ and $(V_1, p_1)_{empty}$, (see Figure 1.b and Figure 1.c):

$$f_\kappa = \left(\frac{\gamma}{\gamma-1}\right)\frac{p_m}{V_{tot}\varphi}\left[\left(\frac{V_1}{p_1}\right)_{full} - \left(\frac{V_1}{p_1}\right)_{empty}\right] + \frac{1}{\gamma-1}\frac{1-\varphi}{\varphi} \tag{3}$$

In the present work, the mechanical input of Hayden's apparatus (oscillating piston) is substituted with an acoustic driver placed between two small cavities and the porous material is not up scaled. For wavelengths much larger than the maximum linear dimension of the cavities, the volume perturbation V_1 is linked to the volume velocity U as follow:

$$V_1 = C_{dw} p_{dw} \tag{4}$$

where p_{dw} is the sound pressure measured in the lower cavity, and $C_{dw} = V_{dw}/\gamma P_0$ is the acoustic compliance of the air in the lower cavity and V_{dw} is the compressible air-volume in the lower cavity.

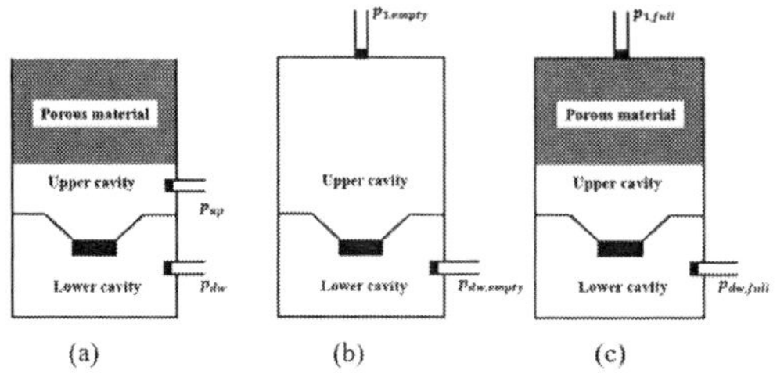

Figure 1. (a) Dragonetti et al. set up for air-flow resistivity measure. (b) "empty" and (c) "full" acoustic set up of Hayden's apparatus.

4 Results and conclusion

Numerical procedure has been used to aim the reliability of the proposed methodology. Different samples have been simulated by fixing their transport parameters. Therefore, dynamic thermal permeability, as well as static thermal one, is obtained combining Eq. (3), (4) in (2). In Figure 2, it can be seen that in the low frequency limit of dynamic thermal permeability, the imaginary part tends to zero value, while the real part tends to the fixed value for the tested samples.

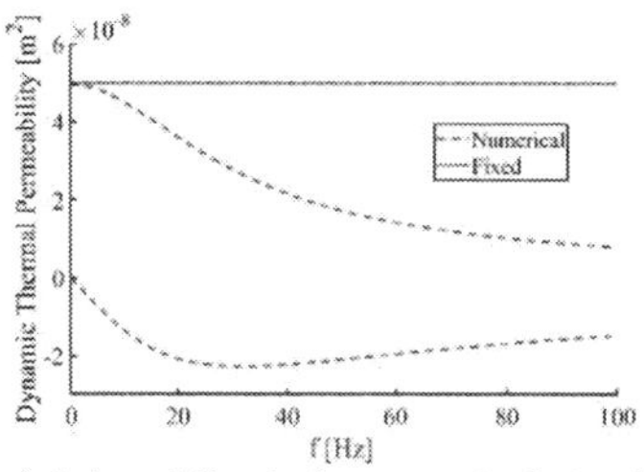

Figure 2. Dynamic thermal permeability: (--) dotted line is for numerical simulation, (-) continuous line is the fixed value.

Numerical simulations are carried out by varying the lower cavity volume and for different materials. In the limit of assuming “lumped parameters”, results shows that measures are not affected by V_{dw}. In the table above, estimating percentual errors versus the increasing frequency are reported for different materials. Future experimental tests will be carried out to establish how thermal dissipation in tube can affect the measures and their repeatability.

$k0'$ fixed [m²]	3.41E-08		1.4411E-08		7.2046E-09	
f [Hz]	$k0'$ evaluated [m²]	%error	$k0'$ evaluated [m²]	%error	$k0'$ evaluated [m²]	%error
2	3.418E-08	-0.34	1.446E-08	-0.31	7.219E-09	-0.20
5	3.408E-08	-0.03	1.445E-08	-0.25	7.218E-09	-0.18
10	3.370E-08	1.08	1.441E-08	-0.01	7.213E-09	-0.12
15	3.309E-08	2.87	1.436E-08	0.37	7.205E-09	-0.01
20	3.227E-08	5.27	1.428E-08	0.90	7.194E-09	0.15

Table 1. Fixed and evaluated data for thermal permeability.

References

[1] D. L. Johnson, J. Koplik, R. Dashen. “Theory of dynamic permeability and tortuosity in fluid-saturated porous media,” J. Fluid Mech. 176, 379–402 (1987).

[2] D. Lafarge, P. Lemarinier, J. F. Allard, V. Tarnow. “Dynamic compressibility of air in porous structures at audible frequencies”. J. Acoust. Soc. Am. 102 (1997) 1995-2006.

[3] J.F. Allard, Y, Champoux. “New empirical equations for sound propagation in rigid frame fibrous materials”. J. Acoust. Soc. Am. 91 (1992), 3346-3353.

[4] R. Dragonetti, C. Ianniello, R.A. Romano. “Measurement of the resistivity of porous materials with an alternating air-flow method”. J. Acoust. Soc. Am. 129 (2), February 2011.

[5] M. E. Hayden, G. W. Swift. “Thermoacoustic relaxation in a pin-array stack”. J. Acoust. Soc. Am. 102 (1997), 2714-2722.

[6] C. Perrot, R. Panneton, X. Only. “Computation of the dynamic thermal dissipation properties of porous media by Brownian motion simulation: Application to an open-cell aluminum foam”. J. Applied Physics 102, 074917 (2007).

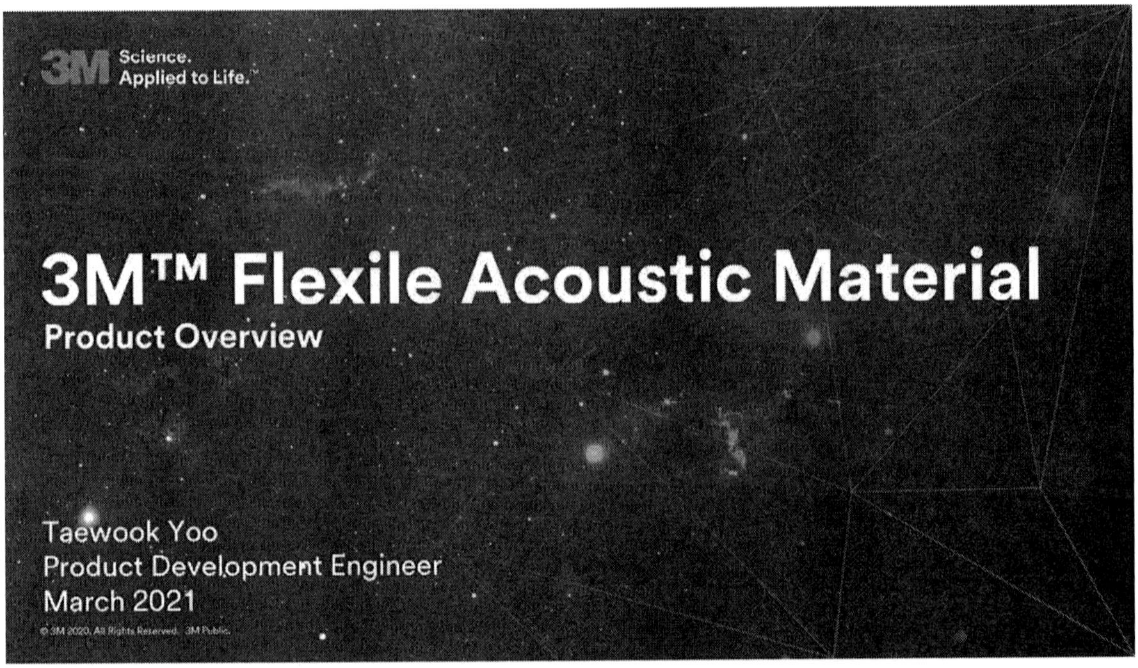
Science.
Applied to Life.™
3M™ Flexile Acoustic Material
Product Overview
Taewook Yoo
Product Development Engineer
March 2021
© 3M 2020. All Rights Reserved. 3M Public.

3M Automotive NVH-T Solutions Roadmap
1990s - 2016
3M™ Thinsulate™ Acoustic Insulation TAI, SM, AU, TC Series
• High absorption performance
• Lightweight
3M™ Thinsulate™ High Temperature Acoustic Absorbers HT Series
• Higher thermal resistance for engine room applications (2016)
2018
Higher Value Absorbers
3M™ Thinsulate™ Acoustic Insulation TF Series
• Higher absorption performance per unit weight/thickness
• Save cost without sacrificing performance
2019
Expand Beyond Absorbers (and into Electric Vehicles)
3M™ Engineered Damping Material
• High damping performance per unit weight
• Flexible material with adhesive backing
3M™ Battery Enhancement Material
• FR rated thermal insulation to improve battery performance in EVs
2020 and beyond
3M™ Flexile Acoustic Material
• High absorption performance (especially at mid frequency)
• Demonstrates transmission loss
© 3M 2021. All Rights Reserved. 3M Public.
3

How Does 3M™ Flexile Acoustic Material Work?

Current Solution

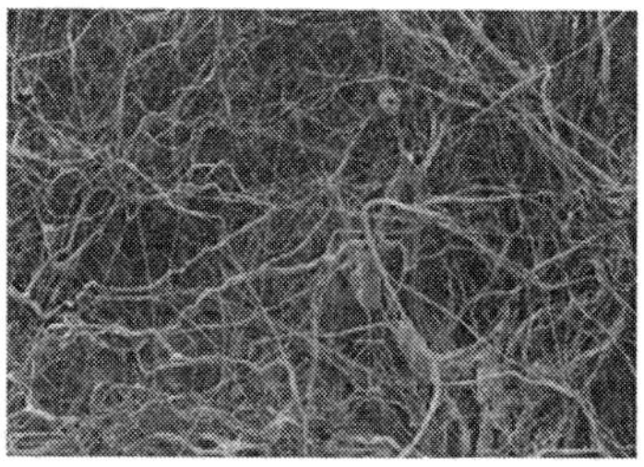

~1 m² per gram

3M FAM

~250 m² per gram

3M™ Flexile Acoustic Material has extremely high specific surface area and high airflow resistivity

Material Test: 3M FAB0910 vs. Typical Barrier Composite

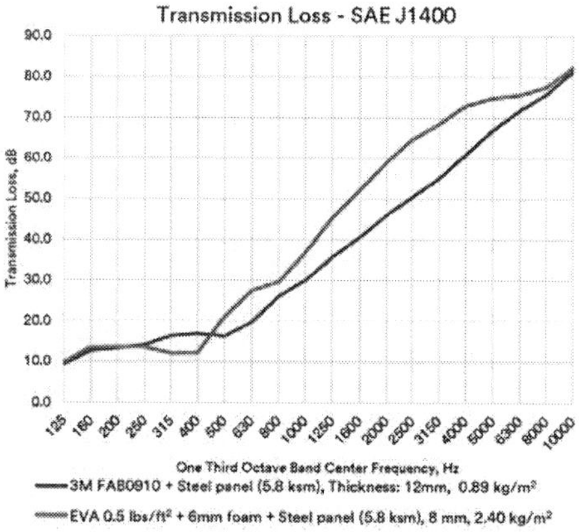

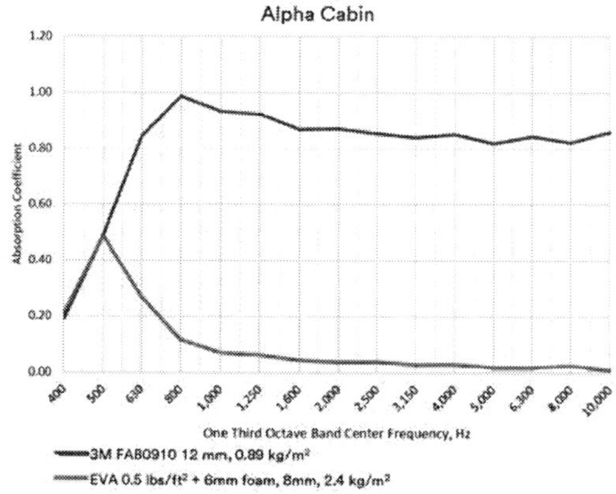

Good barrier and excellent absorber performance at approx. ⅓ the weight of typical barrier composite

In-Vehicle Test Result on Wheelhouse in SUV

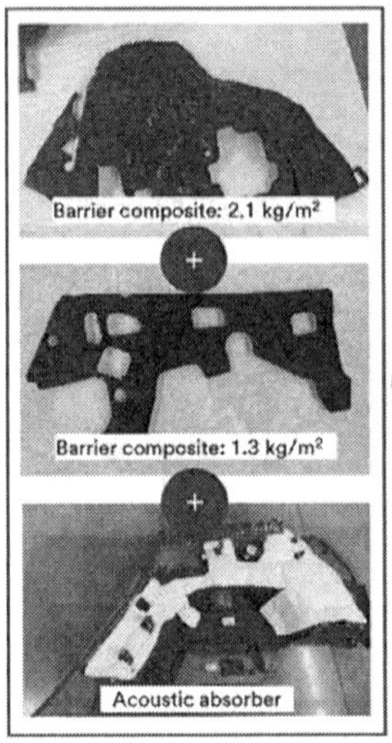

=

3M Flexile Acoustic Material FAB0910, 12mm, 1.2 kg/m²

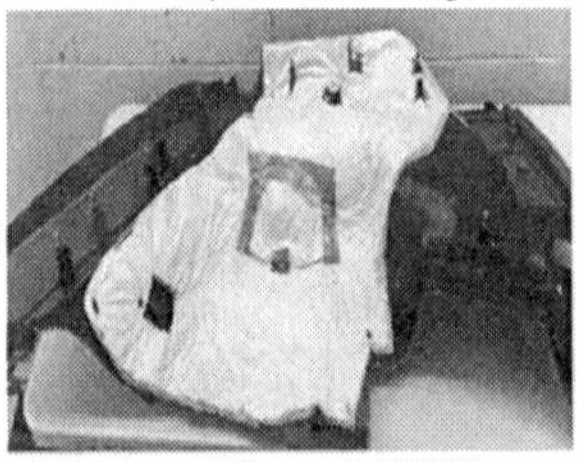

One part of 3M FAB0910 replaces three different parts and has better performance than baseline

SPL over 3 interior mics at 50 mph speed

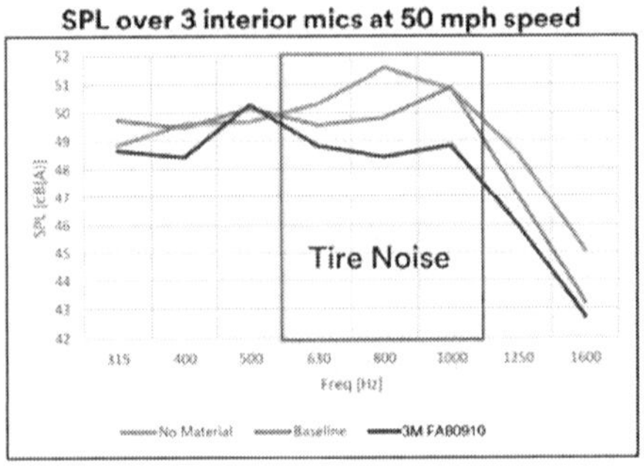

3M

3M FAB0910: Improved Performance, Added Benefits

Absorber and barrier performance in a non-molded single layer	Multifunctional/Simplification, Eliminates molding costs, Potential system cost reduction
Lightweight system solution	Potentially Improves Fuel Efficiency
Equivalent acoustic performance when attached to the body side or the trim side compared to current solutions	Business process re-engineering
Thin product with high acoustic performance	Design flexibility and/or increased cabin space

3M

Thank you!

Important Notice and Disclaimer: This 3M product is an experimental or developmental product that has not been introduced or commercialized for general sale, and its formulation, performance characteristics, and other properties, specifications (if any), availability, and pricing are not guaranteed and are subject to change or withdrawal without notice. User is solely responsible for evaluating third party intellectual property rights and for ensuring that user's use of this 3M product does not violate any third party intellectual property rights.

Technical Information: The technical information, guidance, and other statements contained in this document or otherwise provided by 3M are based upon records, tests, or experience that 3M believes to be reliable, but the accuracy, completeness, and representative nature of such information is not guaranteed. Such information is intended for people with knowledge and technical skills sufficient to assess and apply their own informed judgment to the information. No license under any 3M or third party intellectual property rights is granted or implied with this information.

Product Selection and Use: Many factors beyond 3M's control and uniquely within user's knowledge and control can affect the use and performance of a 3M product in a particular application. As a result, user is solely responsible for evaluating this 3M product and determining whether it is appropriate and suitable for customer's application, including conducting a workplace hazard assessment and reviewing all applicable regulations and standards (*e.g.*, OSHA, ANSI, etc.). Failure to properly evaluate, select, and use a 3M product and appropriate safety products, or to meet all applicable safety regulations, may result in injury, sickness, death, and/or harm to property.

Warranty Disclaimer: This 3M product is sold or made available "AS IS." 3M MAKES NO WARRANTIES OR CONDITIONS, EXPRESS OR IMPLIED, INCLUDING, BUT NOT LIMITED TO, ANY IMPLIED WARRANTY OR CONDITION OF MERCHANTABILITY, FITNESS FOR A PARTICULAR PURPOSE, OR ARISING OUT OF A COURSE OF DEALING, CUSTOM, OR USAGE OF TRADE.

Limitation of Liability: Except to the extent prohibited by law, 3M will not be liable for any loss or damage arising from or related to this 3M product, whether direct, indirect, special, incidental, or consequential (including, but not limited to, lost profits or business opportunity), regardless of the legal or equitable theory asserted, including, but not limited to, warranty, contract, negligence, or strict liability.

Automotive and Aerospace Solutions Division
3M Center
St. Paul, MN 55144-1000
Phone: 1-800-328-1684
Web: www.3M.com/acousticsolutions

MATELYS
RESEARCH LAB

contact@matelys.com

SAPEM 2020+1
6TH TRIENNIAL

SYMPOSIUM ON THE ACOUSTICS OF
PORO-ELASTIC MATERIALS

Presentation outline

- Matelys – general presentation
- How to manage uncertainties and standard deviations from the measurements to the simulations

Matelys > Overview

Matelys is an independent research laboratory which expertise covers materials, acoustics, mechanics, thermodynamics, flow

Accredited French Research Ministry (*CIR*) since 2007
and awarded by peers

AWARD 2011
RESEARCH
GOLD DECIBEL

Industry Award 2012

www.matelys.com | contact@matelys.com

MATELYS
RESEARCH LAB

Matelys > Activities

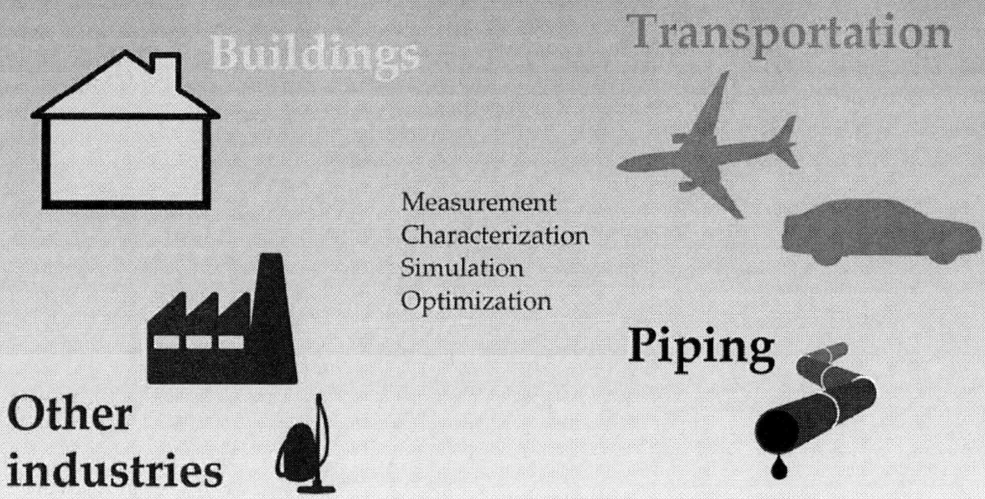

www.matelys.com | contact@matelys.com

MATELYS
RESEARCH LAB

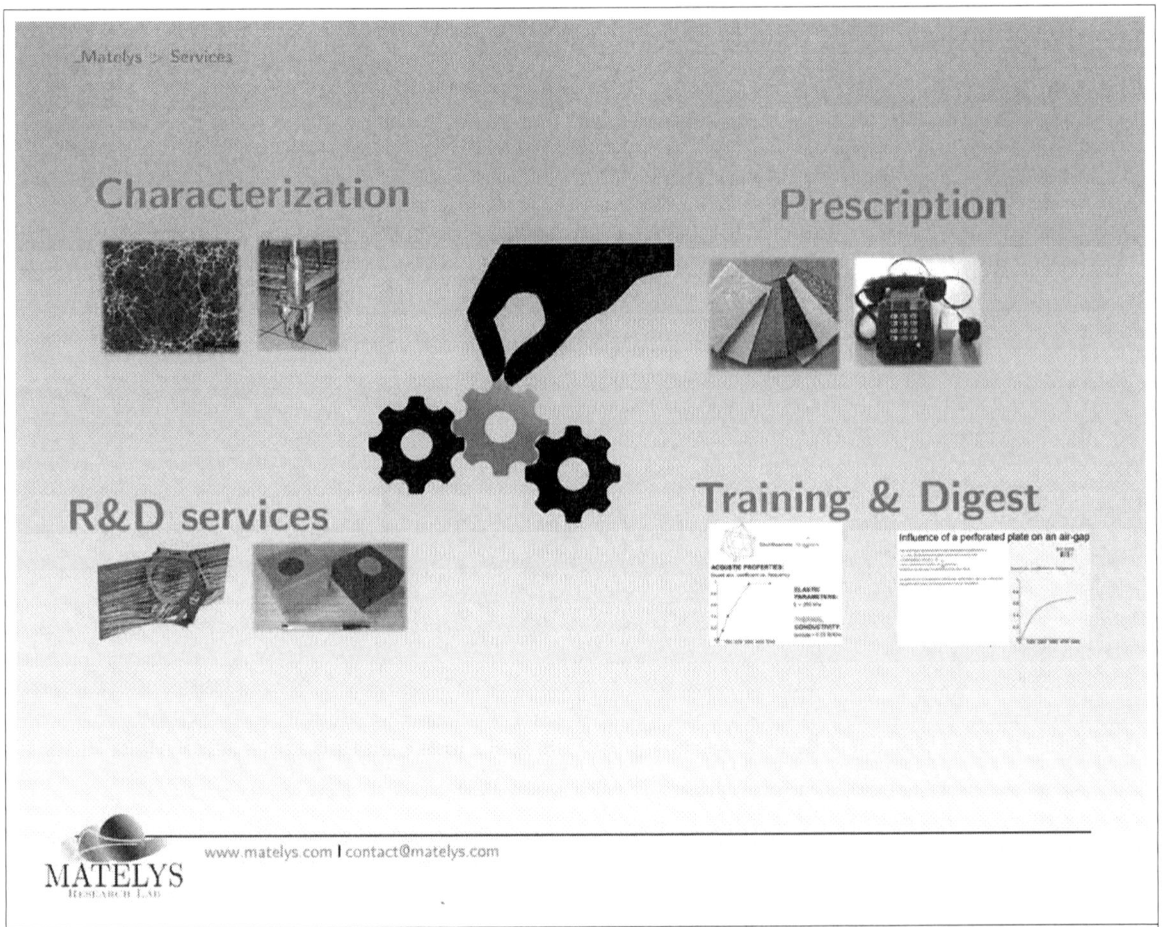

Matelys > Services
Characterization
Prescription
R&D services
Training & Digest
Influence of a perforated plate on an air-gap
MATELYS
RESEARCH LAB
www.matelys.com I contact@matelys.com

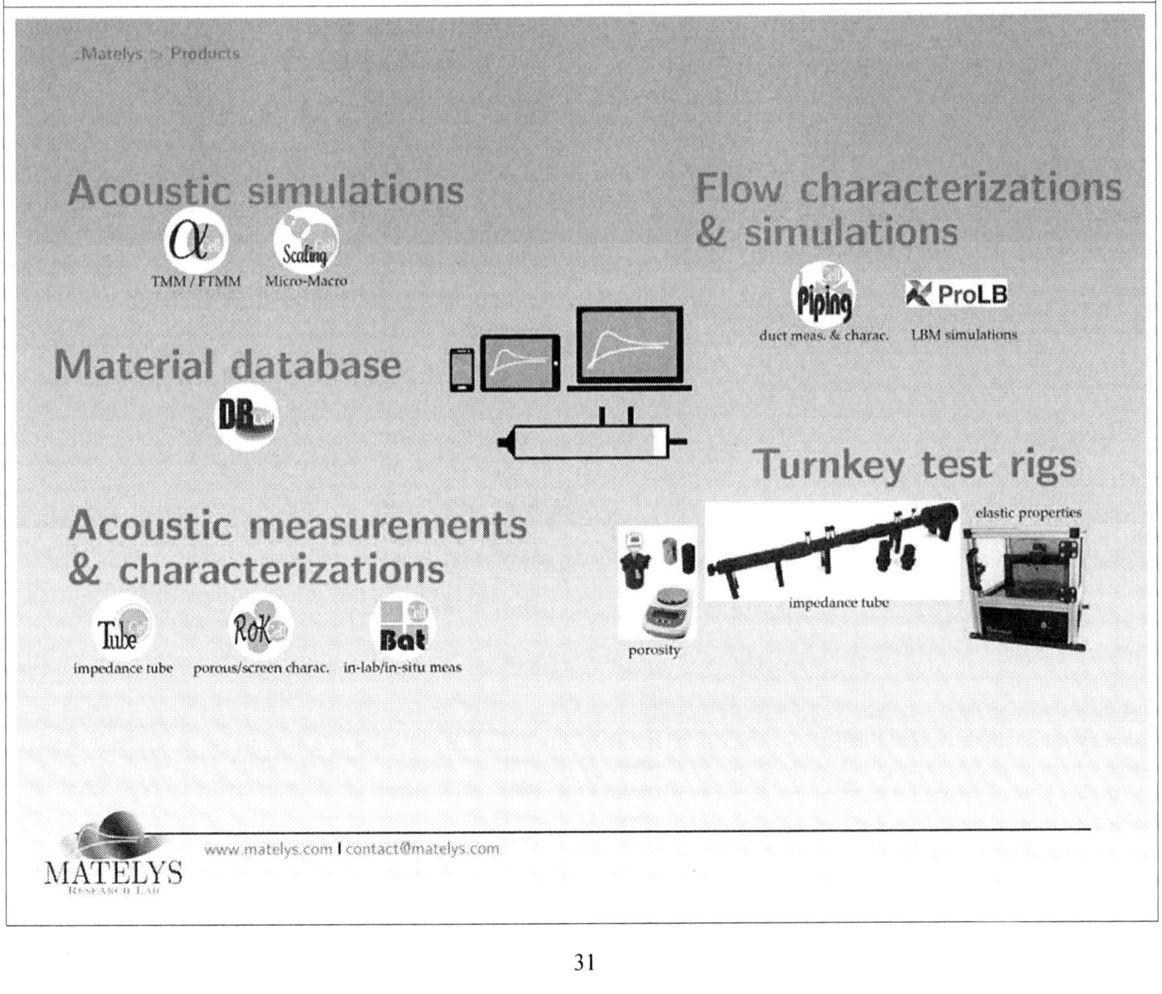

Matelys > Products
Acoustic simulations
TMM / FTMM
Scaling
Micro-Macro
Flow characterizations & simulations
Piping
duct meas. & charac.
ProLB
LBM simulations
Material database
DB
Turnkey test rigs
elastic properties
impedance tube
porosity
Acoustic measurements & characterizations
Tube
impedance tube
RoK
porous/screen charac.
Bat
in-lab/in-situ meas
MATELYS
RESEARCH LAB
www.matelys.com I contact@matelys.com

R&D of common interest

- Material development workflow
- Managing uncertainties and standards deviations from the measurements to the simulations

Matelys in a typical engineering workflow

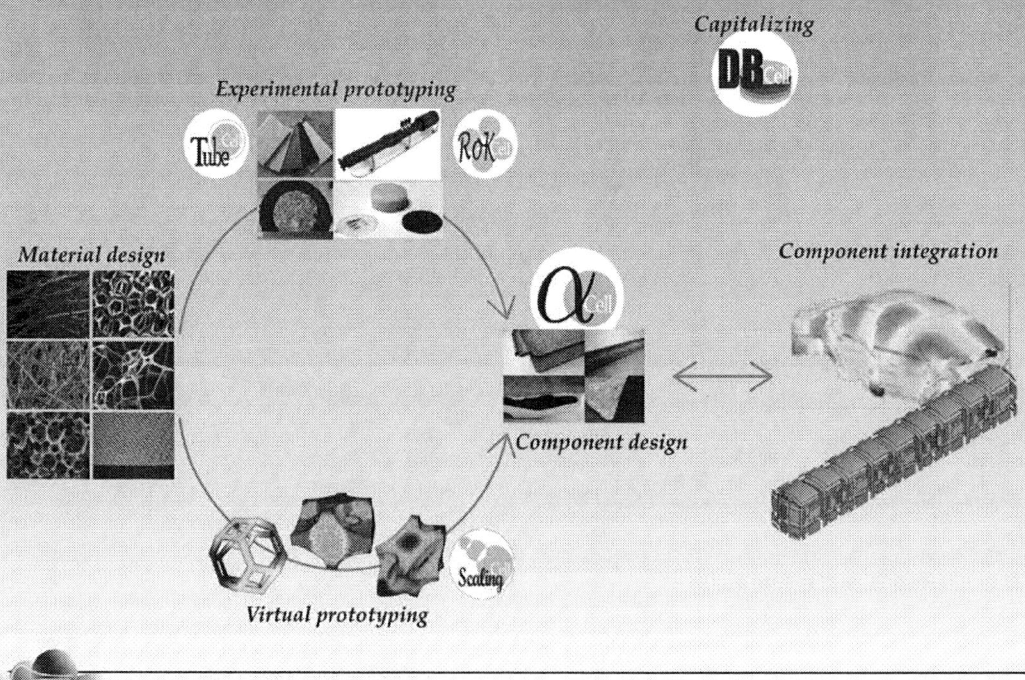

Matelys I > Matelys software suite

Matelys in a typical engineering workflow

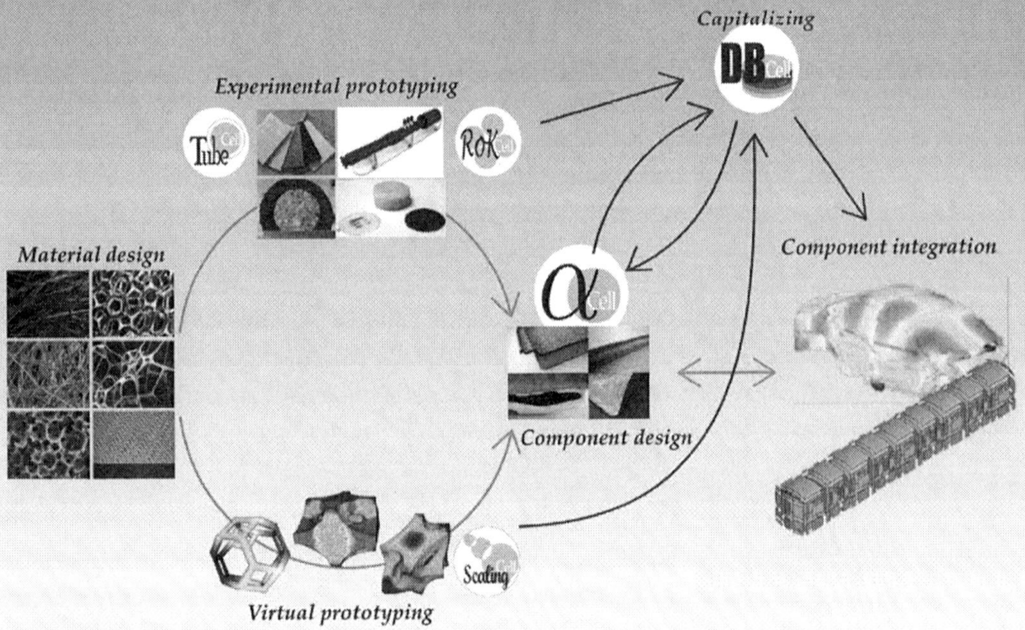

MATELYS
RESEARCH LAB

www.matelys.com I contact@matelys.com

Matelys I > Uncertainties

Managing uncertainties and standard deviations

- Each parameter is measured or characterized with a given **uncertainty**

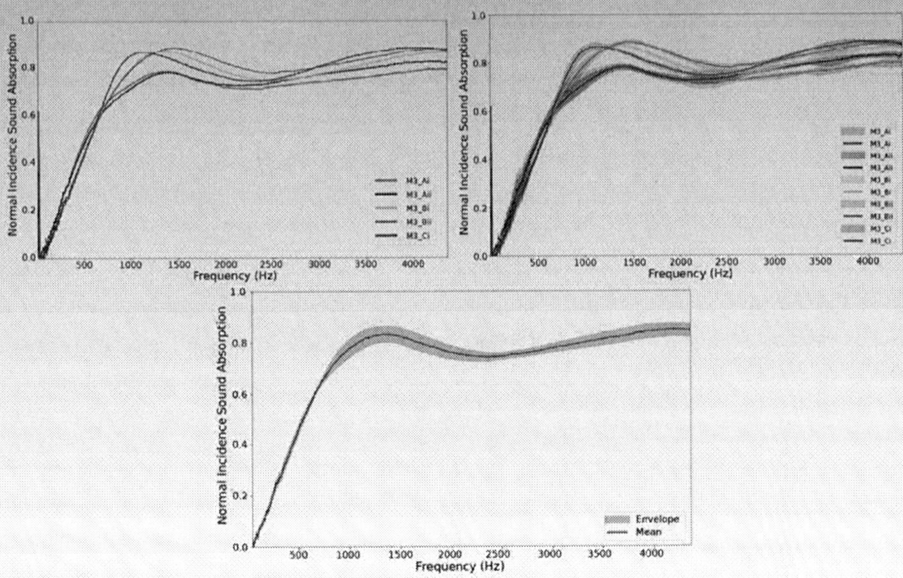

MATELYS
RESEARCH LAB

www.matelys.com I contact@matelys.com

Managing uncertainties and standard deviations

- Each parameter is measured or characterized with a given **uncertainty**
- Several samples are tested to obtain a mean value and a **standard deviation**

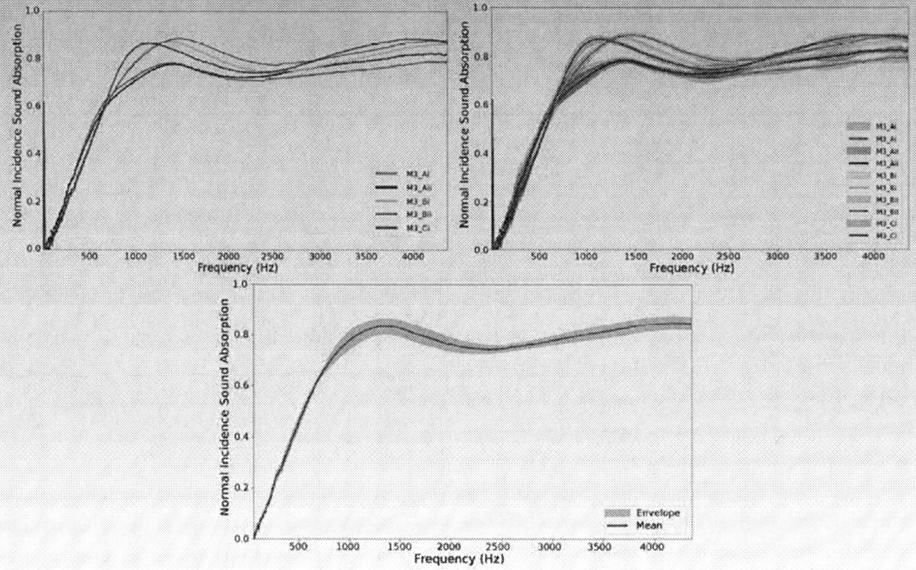

_Matelys | > Uncertainties

Managing uncertainties and standard deviations

- Each parameter is measured or characterized with a given **uncertainty**
- Several samples are tested to obtain a mean value and a **standard deviation**
- These uncertainties and standard deviations are rarely used in simulations

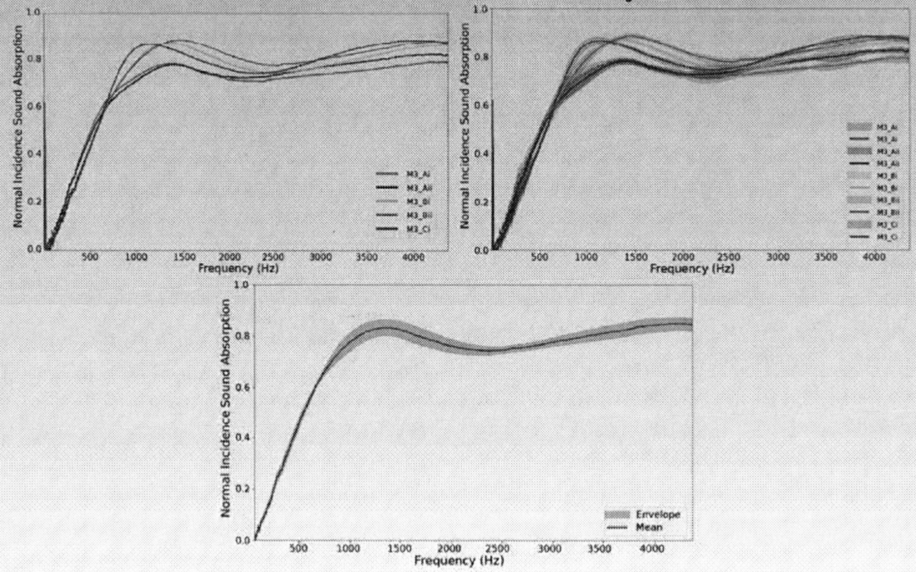

MATELYS
Research Lab
www.matelys.com | contact@matelys.com

Analytical inversion or numerical fitting?

- Two main classes of characterization methods are employed: **analytical inversion** method or **numerical fitting procedure**

_Matelys I > Uncertainties

Analytical inversion or numerical fitting?

- Two main classes of characterization methods are employed: **analytical inversion** method or **numerical fitting procedure**

- How to have confidence in the characterized parameters and their uncertainties?

www.matelys.com I contact@matelys.com
MATELYS Research Lab

Analytical inversion or numerical fitting?

- Two main classes of characterization methods are employed: **analytical inversion** method or **numerical fitting procedure**
- How to have confidence in the characterized parameters and their uncertainties?
- Analytical inverse methods suffer from high sensitivity to the operator.

_Matelys I > Uncertainties

Analytical inversion or numerical fitting?

- Two main classes of characterization methods are employed: **analytical inversion** method or **numerical fitting procedure**
- How to have confidence in the characterized parameters and their uncertainties?
- Analytical inverse methods suffer from high sensitivity to the operator.
- Numerical fitting methods can lead to non physical sets of parameters.

www.matelys.com I contact@matelys.com
MATELYS
RESEARCH LAB

Analytical inversion or numerical fitting?

- Two main classes of characterization methods are employed: **analytical inversion** method or **numerical fitting procedure**
- How to have confidence in the characterized parameters and their uncertainties?
- Analytical inverse methods suffer from high sensitivity to the operator.
- Numerical fitting methods can lead to non physical sets of parameters.
- The fitting procedure can be bounded by the analytical inversion.

_Matelys I > Uncertainties

Analytical inversion or numerical fitting?

- Two main classes of characterization methods are employed: **analytical inversion** method or **numerical fitting procedure**
- How to have confidence in the characterized parameters and their uncertainties?
- Analytical inverse methods suffer from high sensitivity to the operator.
- Numerical fitting methods can lead to non physical sets of parameters.
- The fitting procedure can be bounded by the analytical inversion.
- **Bayesian method** enables to properly couple these methods in order to increase the **confidence in the resulting uncertainties**.

www.matelys.com I contact@matelys.com
MATELYS
RESEARCH LAB

Comparing envelopes

- A **simulated envelope** can be compared to the measured one instead of comparing a single curve!

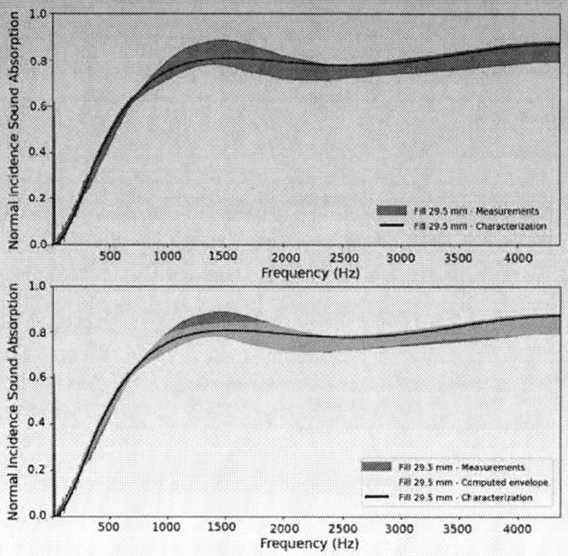

Prepare to be MATELYS approved !

More about porous materials : http://APMR.matelys.com/

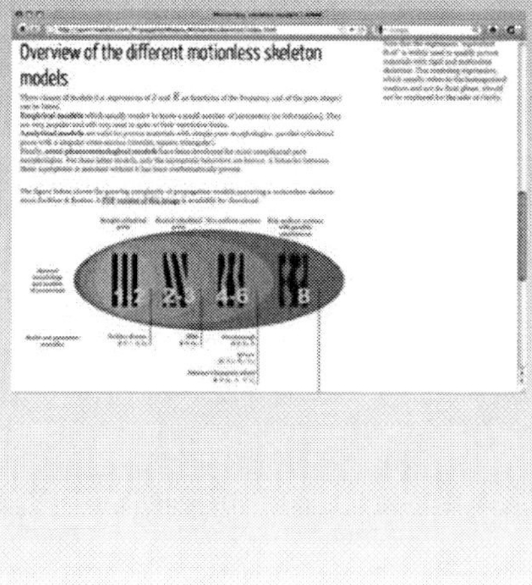

Rapid additive manufacturing of optimised anisotropic metaporous surfaces for broadband absorption

T. Cavalieri[1,2], J. Boulvert[1,2,3], V. Romero-García[1], G. Gabard[1], M. Escouflaire[2], J. Regnard[2], and J.-P. Groby[1]

[1] Laboratoire d'Acoustique de l'Université du Mans (LAUM), UMR CNRS 6613, Institut d'Acoustique - Graduate School (IA-GS), CNRS, Le Mans Université, France

[2] Safran Aircraft Engines, Rond Point René Ravaud - Réau, 77550 Moissy-Cramayel, France

[3] Laboratoire d'Analyse Vibratoire et Acoustique, LAVA, Department of Mechanical Engineering, École Polytechnique de Montréal, P.O. Box 6079 Station Centre-ville, Montréal, Québec H3C 3A7, Canada

Porous materials and metamaterials play a key role in sound absorbing and insulation solutions in acoustics. With the growing interest in additive manufacturing techniques, recent work has focused on the printing of porous and resonant structures for acoustic purposes. Usual metaporous surfaces/interfaces are generally built by periodically inserting resonant elements in an existing porous layer. This complex manufacturing process can be significantly simplified by using additive manufacturing techniques, which also eases the design and optimisation of the metaporous surface. In this work, the acoustic properties of the metaporous surface are controlled by simple geometric parameters defining both the anisotropic porous layer and the shapes of the resonators. Hence, we focus on optimising split-ring resonators embedded in a micro-treillis porous layer, which are built in a single part using additive manufacturing techniques. A finite-element method together with the Bloch wave decomposition provides a numerical model used to predict the reflection and absorption coefficients under normal incidence. The geometric parameters of the anisotropic metaporous surface are then optimised by non-linear minimisation techniques to maximise acoustic absorption. An optimal metaporous surface is 3D printed by fused-deposition modeling and its acoustic properties are measured in an impedance tube. The measurements are in good agreement with the predicted optimal broadband absorption coefficient. This work demonstrates the benefits of additive manufacturing for designing metaporous acoustic surfaces.

1 Optimal design

Porous materials and metamaterials are commonly used in acoustics, for absorption and insulation purposes[1, 2, 4, 3]. In this work, we propose an optimal design of resonant metaporous surface for acoustic absorption purposes. The design of the metaporous surface consists in acoustically rigid 2D split-ring resonators (SRR) periodically embedded in a transverse isotropic porous material, as in Figure 1. By tuning the macro-geometry of rigid resonators as well as the micro-geometry of the porous medium, we maximise the absorption in the frequency range between 1 and 4 kHz for a material thickness of 25 mm. The metaporous surface provides strong broadband absorption capabilities and subwavelength quasi-perfect absorption.

In order to optimise the geometric parameters of the metaporous surface, a constrained non-linear minimisation routine is employed. We jointly optimise the micro- and macro-geometry of the metaporous surface using the Nelder-Mead method [5, 6]. In order to do so, we define a cost function that translates how far the current vector of parameters is from the optimal value $\alpha = 1$. The acoustic absorption coefficient is maximised when the cost function gets closer to zero. The cost function is the arithmetic average of the absorption within $f_{\min} = 1000\,\text{Hz}$ and $f_{\max} = 4000\,\text{Hz}$, with $N_{\text{f}} = 10$, the number of frequencies.

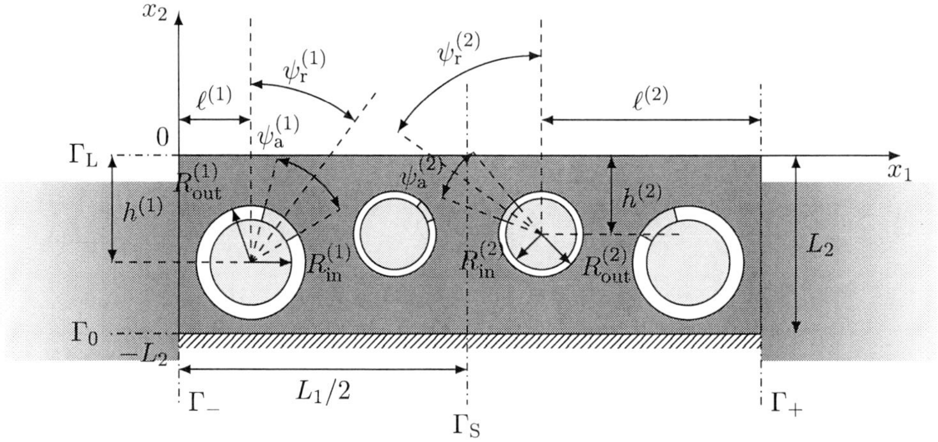

Figure 1: Geometry of the metaporous surface.

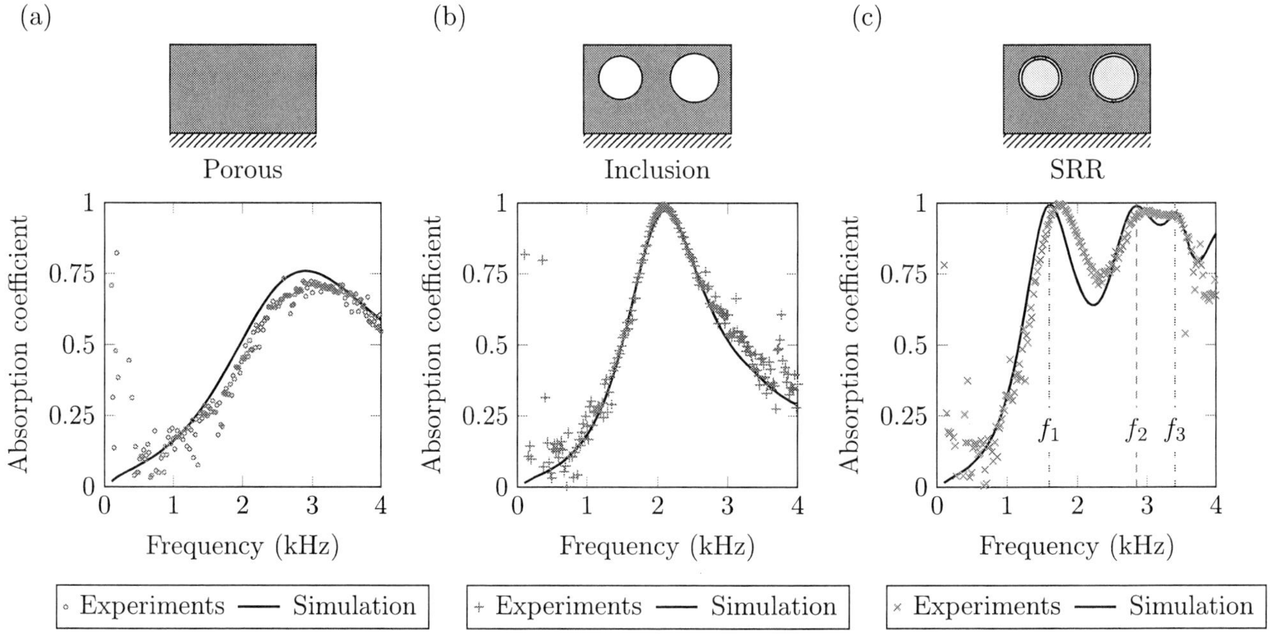

Figure 2: Experimental and predicted results.

2 Conclusions

The numerical simulations based on FEM provide a predictive model for the reflection and absorption coefficients. This predictive model is used in the cost function for the minimisation algorithm, whose goal is to maximise the absorption in the target frequency range. Experiments have been conducted in an impedance tube with a square cross section. They provide a validation of the numerical model, as the experimental data are in good agreement with the simulated results. Additive manufacturing techniques have proved their ability to build resonant metaporous surfaces. The present work demonstrates the capabilities of metaporous surfaces for absorbing sound waves, and highlights the use of additive manufacturing for such acoustic treatments. While the design and manufacturing of materials incorporating inclusions is usually tedious, the solutions presented here are fabricated in a single stage. In addition to offering a solution that is easy and quick to implement, this first prototype was developed at low cost. Finally, the optimisation of the shape and number of inclusions could offer more possibilities regarding the compromise between structural dimensions and absorption capabilities.

References

[1] J. F. Allard, *Propagation of sound in porous media: modelling sound absorbing materials*, Elsevier Applied Science.

[2] T. J. Cox and P. D'Antonio, *Acoustic absorbers and diffusers: theory, design and application*, CRC Press, Taylor & Francis, CRC Press is an imprint of the Taylor & Francis Group, and informa business, third edition ed.

[3] J.-P. Groby, A. Duclos, O. Dazel, L. Boeckx, and W. Lauriks, *Absorption of a rigid frame porous layer with periodic circular inclusions backed by a periodic grating*, The Journal of the Acoustical Society of America, 129 (2011), pp. 3035–3046.

[4] C. Lagarrigue, J. P. Groby, V. Tournat, O. Dazel, and O. Umnova, *Absorption of sound by porous layers with embedded periodic arrays of resonant inclusions*, The Journal of the Acoustical Society of America, 134 (2013), pp. 4670–4680.

[5] J. A. Nelder and R. Mead, *A simplex method for function minimization*, 7, pp. 308–313.

[6] W. H. Press, S. A. Teukolsky, W. T. Vetterling, and B. P. Flannery, *Numerical recipes: the art of scientific computing*, 2007. OCLC: ocn123285342.

3D printed folded porous material for sub-wavelength absorption of sound

J. Boulvert[1,2,3], J. Costa-Baptista[2], E. R. Fotsing[2], T. Cavalieri[1,3], V. Romero-García[1], G. Gabard[1], A. Ross[2], J. Mardjono[3] and J.-P. Groby[2]

[1] Laboratoire d'Acoustique de l'Université du Mans, LAUM - UMR CNRS 6613, Le Mans Université, Avenue Olivier Messiaen, 72085 Le Mans Cedex 9, France
[2] Laboratoire d'Analyse Vibratoire et Acoustique, LAVA, Mechanical engineering, Polytechnique Montréal, Montréal, Québec, Canada
[3] Safran Aircraft Engines, Villaroche, Rond Point René Ravaud - Réau, 77550 Moisy-Cramayel Cedex,France

Open-cell acoustic materials are widely employed for broadband noise absorption of medium and high frequencies. However, due to their intrinsic loss mechanisms, porous materials suffer from a lack of efficiency in their sub-wavelength regime defined by $\lambda < 4L$ *where* λ *is the wavelength in air and L the thickness of the treatment. A slab of porous material presents low absorption below its so called "quarter wavelength resonance frequency", which is related to the thickness of the slab. Increasing the thickness of a porous treatment can enhance its low frequencies absorption but makes it bulkier and heavier which is problematic in many practical applications. In this work, a metaporous surface is introduced. Its is a treatment made of folded porous materials having an acoustic effective thickness greater than the thickness of the treatment itself. The metaporous surface is optimized, 3D printed and tested experimentally. The mean value of the measured absorption coefficient is 0.98 for frequencies corresponding to a wavelength in air* λ *such that* $5.3L < \lambda < 9.9L$, *where L is the thickness of the treatment.*

1 Presentation of the Metaporous Surface

A metaporous surface (MpS) composed of folded porous materials (FPMs) is designed, optimized for sub-wavelength and broadband absorption of sound, and tested experimentally as reported in [1].

A MpS usually consists in inclusions embedded in a porous material slab [3]. The inclusions increase the density of state in the sub-wavelength regime of the porous slab. The opposite approach is used in this work: folded quarter wavelength resonators are filled by a porous material. The FPMs consist in cavities folded in helical shape and filled with a structured porous medium. Folded cavities have an higher acoustic effective thickness than their actual height. A parallel assembly of several FPMs, each one tuned at a distinct frequency, generates the MpS, see Fig. 1(a).

2 Modeling and optimization

The normal incidence acoustic behavior of the treatment is predicted analytically by a mode matching model [2] and numerically by a Finite Element Method (FEM) model. The mode matching model is used during the treatment optimization and is coupled to a Nelder-Mead optimization algorithm [4] while the FEM model confirms the validity of the analytic predictions.

The effective thickness and the intrinsic losses of the MpS are controlled separately by tunning the pitch of the cavities and the pore size of the filling porous medium. The optimization of the pitch and of the pore size enables to achieve perfect absorption at frequencies much lower than the quarter wavelength frequency of the treatment.

3 Results

A 30 mm thick MpS is optimized to maximize its absorption coefficient in normal incidence and with a rigid backing in [1150; 2000] Hz. It is 3D printed by means of Fused Deposition Modeling technique and tested in an impedance tube. The corresponding absorption coefficients computed analytically and

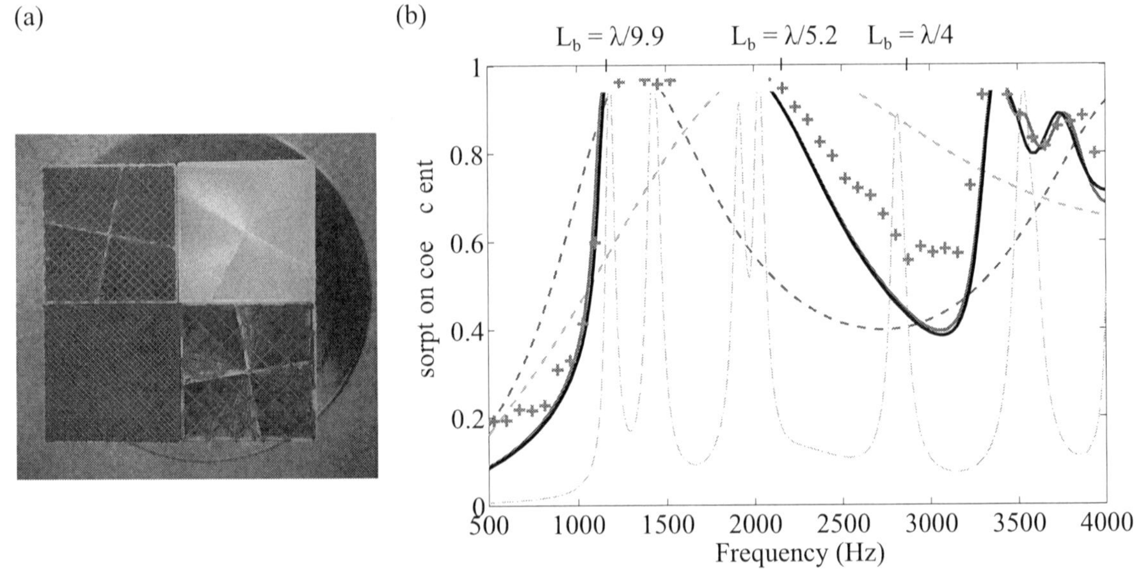

Figure 1: (a) Picture of optimized and 3D printed MpS placed on the rigid termination of the impedance tube.
(b) Absorption coefficient of $L_b = 30$ mm thick materials in normal incidence with rigid backing. Critically coupled: homogeneous porous material (dashed blue line) and single FPM (dashed purple line). Simulation of the optimized MpS: analytical model (solid green line), FEM model (solid black line). Measurement of optimized MpS (red crosses). Analytical simulation of the optimized MpS replacing the porous material by air (dash-doted green line).

numerically and measured experimentally are presented in Fig. 1(b). Simultaneous sub-wavelength and large band absorption are observed: the mean value of the measured absorption coefficient is 0.98 for frequencies corresponding to a wavelength in air λ such that $5.3L < \lambda < 9.9L$, where L is the thickness of the treatment.

Comparison of the MpS absorption to that of (1) straight porous material (blue dashed line), (2) single FpM (purple dashed line) and (3) MpS replacing the filling porous medium by air (yellow dashed line) highlights that (1) folding the cavities allows to reach perfect absorption at lower frequencies, (2) that combining several FpMs tuned at different frequencies allows to broaden the absorption plateau and (3) that the filling porous material controls the losses of the folded cavities leading to the critical coupling condition [5] and decreases the modes quality factor.

References

[1] J. Boulvert, J. Costa-Baptista, T. Cavalieri, V. Romero-García, G. Gabard, E. R. Fotsing, A. Ross, M. Perna, J. Mardjono, and J.-P. Groby. Folded metaporous material for sub-wavelength and broadband perfect sound absorption. *Applied Physics Letters*, 117(25):251902, December 2020.

[2] J.-P. Groby, W. Lauriks, and T. E. Vigran. Total absorption peak by use of a rigid frame porous layer backed by a rigid multi-irregularities grating. *The Journal of the Acoustical Society of America*, 127(5):2865–2874, May 2010.

[3] L. Lagarrigue, J.-P. Groby, V. Tournat, O. Dazel, and B. Nennig. Acoustic panel.

[4] J. A. Nelder and R. Mead. A Simplex Method for Function Minimization. *The Computer Journal*, 7(4):308–313, January 1965.

[5] V. Romero-García, G. Theocharis, O. Richoux, and V. Pagneux. Use of complex frequency plane to design broadband and sub-wavelength absorbers. *The Journal of the Acoustical Society of America*, 139(6):3395–3403, June 2016.

3D PRINTED PLA-HYDROGEL FRACTAL ACOUSTIC COMPOSITE METAMATERIAL

G. Comandini[1], F. Scarpa[1,2], M. Azarpeyvand[2] and V. P. Ting[1,3]

[1]Bristol Composites Institute (ACCIS), University of Bristol,
[2]Department of Aerospace Engineering, University of Bristol,
[3]Department of Mechanical Engineering. University of Bristol,
Queen's Building, University Walk, Bristol BS8 1TR, UK

1 Abstract

In this work COMSOL Multiphysics simulations and experiments were performed on metamaterials with fractal geometries and hydrogel fillers. The results show key features which affect the meta-behaviour of these structures. The metamaterials have a Hilbert fractal geometry (from zero to the third order) with internal gap widths ranging from 0.5 mm to 5 mm. The comparison between metamaterials was carried out by evaluating their response in term of absorption coefficient (AC) and transmission loss (TL) between the empty metamaterial made by polylactic acid (PLA) and the composite one made by hydrogel and PLA.

2 Introduction

Excess noise has detrimental effects on human health and wellbeing. The mitigation of this type of pollution is not trivial but essential to improve the collective quality of life [1]. For mechanical, civil, and aerospace engineering applications, one of the common ways to deal with unwanted sound is to exploit the mass law or utilise thick layers of absorbing materials [2]. However, that is not always possible due to weight or space restraints, and new solutions and technologies are needed. Composite metamaterials can be part of the answer [3, 4].

3 PLA-Hydrogel Composite

Composite materials have many potential applications in several engineering fields. One side which has not yet been explored is the use of composites to produce multiphase metamaterials for acoustic applications. Combining additive manufacturing technology with hydrogel is something new [5], and according to the experimental results they seem quite promising [6]. PLA fractal metamaterials' acoustic performance with and without hydrogel fillers was measured using ASTM standard E2611-09 tests with an impedance tube.

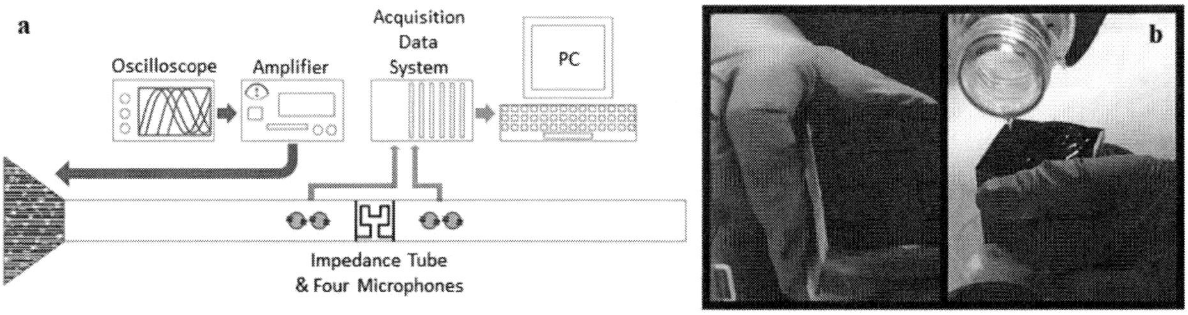

Figure 1: Impedance tube (a), Composite metamaterial (CMMs) manufacturing with hydrogel (b)

Figure 1(a) shows a drawing of the test rig configuration used in the experiments to identify the absorption coefficient (AC) and transmission loss (TL). The specimens were subjected to white noise in the test room. Four microphones linked to an acquisition data system and a PC were used to record data in a range between 200 and 3000Hz. Figure 1(b) shows one of the samples representing the composite metamaterial. The hydrogel used to cast into the metamaterial is an F127 alginate hybrid gel with distilled water. This fluid type was used as an infill material because of its microporous structure and interesting acoustic properties which have not been thoroughly investigated yet [6].

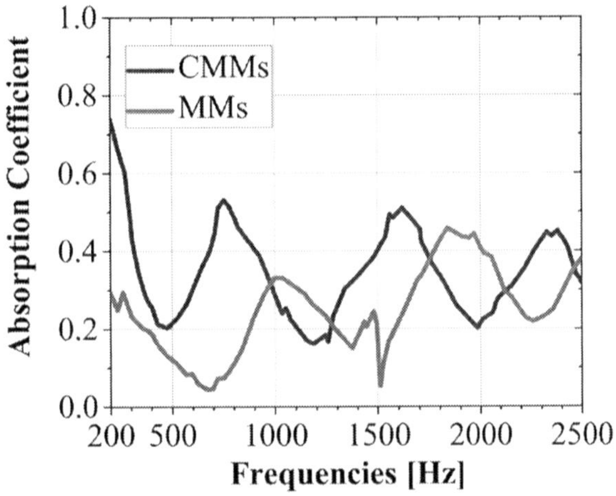

Figure 2: Experimental comparison between CMMs and MMs

Figure 2 shows the difference in terms of absorption coefficient between the composite metamaterial (CMMs) and the traditional metamaterial (MMs). For the low-frequency range (below 500 Hz), the acoustic absorption for the composite metamaterial is more than double compared to the traditional MMs. A high level of AC in some areas of the frequency spectrum (especially at 200 Hz with AC values close to 0.8) indicates a high dissipation of acoustic energy. These results underline a novel way to absorb energy from sound using these composite metamaterials passively.

This work provides an experimental feasibility study on how the introduction of composite metamaterials can contribute to the dissipation of acoustic energy and increase the TL in a structure. Hilbert fractal lattice configurations have been used as a recursive and geometrical shape with different gap widths in this work. Moreover, mechanical characterisation was also performed via compression tests to understand some critical mechanical properties as a step towards practical applications.

References

[1] Kim, Jihyun, Wanhyung Lee, Jong-Uk Won, Jin-Ha Yoon, Hongdeok Seok, Yeong-Kwang Kim, Seunghyun Lee, and Jaehoon Roh. "The relationship between occupational noise and vibration exposure and headache/eyestrain, based on the fourth Korean Working Condition Survey (KWCS)." *PloS one* 12, no. 5 (2017): e0177846.

[2] Ver, Istvan L., and Leo L. Beranek. "Noise and vibration control engineering." *Principles and Applications, (Second Edition), New York, NY, USA* (2008).

[3] Lincoln, Reece L., Fabrizio Scarpa, Valeska P. Ting, and Richard S. Trask. "Multifunctional composites: A metamaterial perspective." *Multifunctional Materials* 2, no. 4 (2019): 043001.

[4] Cummer, Steven A., Johan Christensen, and Andrea Alù. "Controlling sound with acoustic metamaterials." *Nature Reviews Materials* 1, no. 3 (2016): 1-13.

[5] Usta, Fatih, Fabrizio Scarpa, Halit S. Türkmen, Peter Johnson, Adam W. Perriman, and Yanyu Chen. "Multiphase lattice metamaterials with enhanced mechanical performance." *Smart Materials and Structures* 30, no. 2 (2021): 025014.

[6] Zhang, Kai, Chu Ma, Qi He, Shaoting Lin, Yi Chen, Yu Zhang, Nicholas X. Fang, and Xuanhe Zhao. "Hydrogels: Metagel with Broadband Tunable Acoustic Properties Over Air–Water–Solid Ranges (Adv. Funct. Mater. 38/2019)." *Advanced Functional Materials* 29, no. 38 (2019): 1970264.

Natural sonic crystal absorber constituted of Aegagropilae fiber network

L. Barguet[1], V. Romero-García[1], N. Jiménez[2], L.M. Garcia-Raffi [3], V. J. Sánchez-Morcillo [4], and J.-P. Groby[1]

[1] Laboratoire d'Acoustique de l'Université du Mans (LAUM), UMR CNRS 6613, Institut d'Acoustique - Graduate School (IA-GS), CNRS, Le Mans Université, France
[2] Instituto de Instrumentación para Imagen Molecular, Consejo Superior de Investigaciones Científicas, Universitat Politècnica de València, Spain
[3] Instituto de Matemática Pura y Aplicada (IUMPA), Universitat Politècnica de València, Spain
[4] Instituto de Investigación para la Gestión Integrada de Zonas Costeras, Universitat Politècnica de València, Gandia, Spain

We present a 3-dimensional fully natural sonic crystal composed of Aegagropilae that are spherical aggregates resulting from the decomposition of Posidonia Oceanica. The fiber network is first acoustically characterized, providing insights on this natural fiber entanglement due to turbulent flow. The seagrass fibrous spheres are then arranged on a principal cubic lattice. Band diagram and topology of this structure are analyzed, notably via Argand representation of its scattering elements. This fully natural sonic crystal exhibits excellent sound absorbing properties and thus represents a sustainable alternative that could outperform conventional acoustic materials.

1 Introduction

Complex biological structures usually result from the adaption of living bodies to their environmental constraints. These structures generally ensure several functionalities in nature [1]. Very few of them are the result of natural dynamic processes involving organic waste, apparently formed for no particular reason and without ensuring any specific functionality. Aegagropilae such as Posidonia balls are the archetype of these organic structures. They are natural spherical aggregates formed by entangled fibers as a result of the decomposition of Posidonia Oceanica.

2 Fully natural 3-dimensional sonic crystal

Aegagropilae spheres were cut into four parts, resulting in quarter-spheres. Six Aegagropilae quarter spheres of radii 20 ± 1 mm are placed in a square cross-sectional impedance tube with side $d/2 = 21.5$ mm, their centers being separated by a distance $d = 43$ mm. A finite depth perfectly periodic structure is created, composed of six layers incorporating spheres arranged over a square lattice, thus mimicking a finite depth primitive cubic system with a filling fraction $ff \approx 0.4$. The reflection and transmission coefficients are measured via 4 flush mounted microphones for frequencies below the cut-off frequency of the impedance tube.
Figure 1(a) depicts the absolute value of the reflection and transmission coefficients, $|R|$ and $|T|$, as well as the absorption coefficient $\alpha = 1 - |R|^2 - |T|^2$ measured experimentally and calculated by the Finite Element Method. Both measured and simulated coefficients are in good agreement assuming sphere radii of 19.5 mm and fiber network properties as determined previously. The Bragg interference is clearly noticed around $f_B \approx c_0/2d \leq 3950$ Hz, as highlighted by the grey regions. While Bragg interference is usually associated with a band gap, preventing the propagation of acoustic wave, therefore leading to a minimum transmission, the weak impedance contrast between the soft Aegagropilae spheres and the air medium almost prevents the band gap translation in the transmission coefficient. The quasi absence of band gap is also confirmed by the dispersion relation depicted in Fig. 1(b-c). In addition, the reflection coefficient presents five smooth peaks (and six minima) before the Bragg frequency corresponding to the Fabry-Perot interference arising from the six layers (see Fig. 1). These Fabry-Perot interferences are not visible in the transmission coefficient absolute value because of the attenuation.
To get further insights on the acoustic behavior of this primitive cubic system, the Argand diagram [2] of R and T, i.e., their values in the complex plane in function of frequency, are depicted in Fig. 1(d).

Of particular interest is that R describes half a loop from the positive to the negative real half spaces within the first band gap as a translation of the topological character of the gap. Within the second bulk band, R goes back from negative to positive real half spaces around 5500 Hz as a manifestation of the symmetry inversion of the system [3]. This band is thus a nontrivial bulk band with two opposite edge states. A quasi-perfect absorption peak is also noticed around the symmetry inversion frequency at 5500 Hz. The absorption coefficient is higher than that of a full $6 \times d$-thick layer occupied by the fibrous materials as can be seen in Fig. 1(a) for most of the frequency range considered.
More information can be found in Ref. [4].

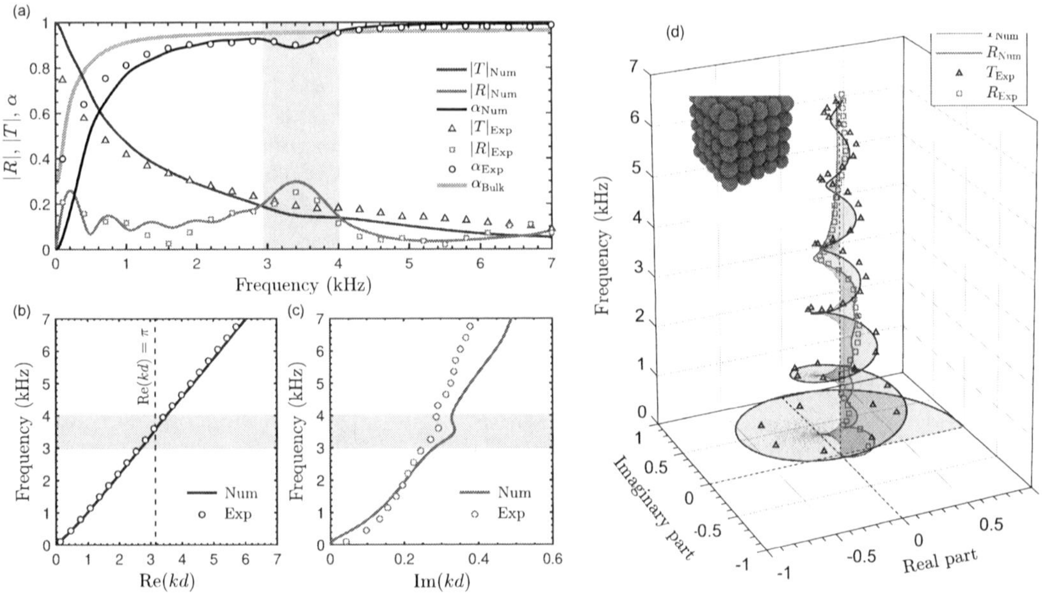

Figure 1: Scattering properties of a 3D primitive cubic arrangement of periodicity $d = 43$ mm, made of $N = 6$ Aegagropilae spheres of radius 19.5 mm. (a) Absorption, reflection and transmission coefficients calculated numerically (black, red and blue continuous curves) and measured experimentally (open black circles, open red squares, and open blue triangles). The grey continuous line represents the absorption coefficient of a bulk material of identical thickness. (b) Real and (c) imaginary parts of the complex wave number reconstructed numerically (continuous lines) and experimentally (markers). (d) Argand diagram of T (blue curve) and R (red curve) over the frequency range [0 Hz,7000 Hz] obtained numerically (continuous) and experimentally (markers).

Acknowledgements

This article is based upon work from COST Action DENORMS CA15125, supported by COST(European Cooperation in Science and Technology). The authors gratefully acknowledge the ANR-RGC META-Room (ANR-18-CE08-0021) project, the project HYPERMETA funded under the program toiles Montantes of the Rgion Pays de la Loire.

References

[1] M. Eder, S. Amini, and P. Fratzl. Biological composites—complex structures for functional diversity. *Science*, 362(6414):543–547, 2018.

[2] A. A. Fernández-Marín, N. Jiménez, J.-P. Groby, J. Sánchez-Dehesa, and V. Romero-García. Aerogel-based metasurfaces for perfect acoustic energy absorption. *Applied Physics Letters*, 115(6):061901, 2019.

[3] M. Xiao, Z. Q. Zhang, and C. T. Chan. Surface impedance and bulk band geometric phases in one-dimensional systems. *Physcial Review X*, 4:021017, 2014.

[4] L. Barguet, V. Romero-García, N. Jiménez, L. M. Garcia-Raffi, V. M. Sánchez-Morcillo, and J.-P. Groby. Natural sonic crystal absorber constituted of seagrass (posidonia oceanica) fibrous spheres. *Sci. Rep.*, 11:711, 2021.

Sound insulation of clamped single panel and infinite double panel system by addition of poro-elastic resonators

Ke Li[1], N. Dauchez[1], B. Nennig[2]

[1] Université de technologie de Compiègne, CNRS, Roberval, Centre de recherche Royallieu - CS 60 319 - 60 203 Compiègne Cedex, France
[2] Institut supérieur de mécanique de Paris (SUPMECA), Laboratoire Quartz EA 7393, 3 rue Fernand Hainaut, 93407 Saint-Ouen, France.

Poroelastic lamellar metamaterial is proposed to face the sound transmission loss dip of the double panel wall. The metamaterial is a network of compressed melamine foam beams clamped to the panel. First, measurements of the sound transmission loss of a clamped single panel in a rectangular duct is performed. It is shown that the lamellar metamaterial increases significantly the sound insulation around the first bending resonance of the panel between 200 Hz and 400 Hz. Several sizes of poroelastic beams allows to spread the effect on a larger frequency range. Dissipation mechanisms are further studied by mean of a finite element model: structural and viscous dissipations dominate around the first beam resonance frequency while thermal dissipation increases slowly with frequency. The concept is then studied on a infinite size double panel configuration by mean of a periodic finite element model. The resonance of the metamaterial is tuned to the double wall resonance. The influence of some parameters like poroelastic material density and incident angle is discussed. To conclude, this work shows that the proposed concept is an effective way to avoid the sound transmission loss dip of a finite-size panel and a double partition in the low frequency range without adding discrete and heavy resonant systems. Moreover the poroelastic nature of the metamaterial allows to take advantage of both resonant and sound absorption properties.

Maximizing sound absorbing performance of PU foam by the layered GO nanoplatelets loading

Inhwa Jung [1], Jaehyung Kim [1], Juhyun Jeon[2], Yeon June Kang [2]

[1]Kyung Hee University, Department of Mechanical Engineering,
Yongin, 17104, Republic of Korea
[2] Seoul National University, School of Mechanical and Aerospace Engineering,
Seoul, 151-744, Republic of Korea

The sound absorption performance of graphene oxide (GO) impregnated polyurethane (PU) foam was studied. PU foams are fabricated to have a wide range of GO loading density by step-by-step impregnation method. It was determined that the flow resistance and the sound absorption coefficient were increased as a function of GO loading, and other foam properties as well. In addition to impregnating GO with a simple structure, a method for optimizing the spatial arrangement of GO impregnated PU was presented. The impregnation structure with the optimal sound absorption rate was suggested by controlling the impregnation structure in the multilayer structure of the porous polyurethane sponge. To this end, we developed a computer algorithm to find the optimal sound absorption structure. Through the simulation, the average sound absorption coefficient between 500 Hz to 4 kHz was maximized as 0.91 by only 56mm thick PU foam, which was verified by the impedance tube measurement. Also, various possibilities were revealed through extensive experiments and simulations.

1 Introduction

As the importance of noise control increases, there is a continuous need for more effective sound absorbing materials. [1] Although all materials can absorb sound as a passive medium, the materials with high sound absorption capability among these materials are very limited. The most efficient sound absorbing materials are mainly porous or fibrous, consisting of open, semi-open or closed pore structures interconnected with regular or irregular shapes. [2]

In general, porous materials have structural characteristics with channels through which sound waves can enter the material. The principle of sound dissipation is that energy is lost due to heat loss due to friction between the air molecules and the pore walls and viscosity loss due to the viscosity of the air flow in the material. This energy consumption principle gives the porous material a wide frequency band for sound absorption. [3,4] The porous structure is a suitable form for dissipating this energy. In particular, it has the advantage of being relatively easy to manufacture and light due to its low density. [5]

Among various porous materials, polyurethane foam (PU) is the most widely used material due to its effective acoustic damping and low cost. In addition, polyurethane is a material that can be manufactured in a variety of soft foams, high elastic foams, or other porous materials, as required. The sound waves penetrate into the pores of the porous PU and are dissipated by viscous friction and heat exchange occurring in the interconnected pores inside. However, in the open cell type PU foam, it is known that it is very difficult to obtain excellent sound absorption performance in a wide frequency range when the sound absorbing material is thin. To overcome this, optimization of the sound absorbing capability has been attempted through optimization of the porous structure itself. [6] Meanwhile, a method of mixing PU with other materials has been attempted. Various optimization methods have been added to the method of improving the sound absorption performance by mixing the material with the PU, thus providing a basis for maximizing the sound absorption performance. [7, 5] Recently, nanomaterials have been noted as a material to be mixed to improve the sound absorption capability of the porous material. In particular, additives such as graphene oxide (GO) and carbon nanotubes (CNT) have been found to be very effective. [8-14]

In this study, graphene oxide (GO) was added into PU foam to maximize sound absorption capability. Especially, the layer optimization method was applied. In particular, through the optimal arrangement of the layers, it was possible to noticeably improve the sound absorption performance in the low frequency range. As a result, the GO impregnated multi-layer PU foam system can have excellent sound absorption coefficient over the entire frequency range, thus industrial application of the result is expected in near future.

2 Theory and Strategy

2.1 Strategy for maximizing sound absorbing coefficient by multi-layer arrangement

To optimize the sound absorption coefficient, the layers were divided into n layers, and each layer was able to combine sound absorbing materials impregnated with GO up to 6 times. Multi-layer sound absorbing material absorption coefficient calculation is performed by a MATLAB code constructed based on a matrix model from references. [15, 16] The optimization routine was coded separately so that the combination with the maximum average sound absorption coefficient could be selected. The average sound absorption coefficient was obtained through the following equation.

$$\bar{\alpha}=\frac{\alpha_{500\,Hz}+\alpha_{1\,kHz}+\alpha_{2\,kHz}+\alpha_{4\,kHz}}{4}$$

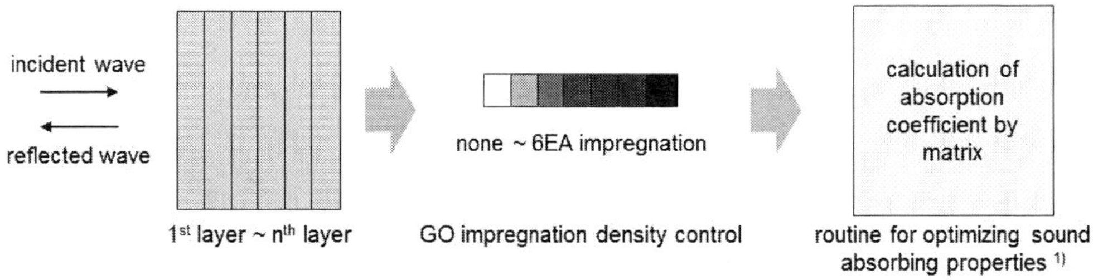

Figure 1: Strategy for maximizing sound absorbing coefficient by multi-layer arrangement

2.2 Impregnation of graphene oxide into PU foam

To verify the calculated sound absorption coefficient experimentally, thin layers of GO impregnated PU foam were prepared. To impregnated GO into PU foam, vacuum impregnation method was applied. The thickness of each layer was set to 4 mm. Therefore, the thickness of the eight layers corresponds to 32 mm. Since each layer is impregnated with GO in 7 steps (0-6 times), a total of $7^8 = 5{,}764{,}801$ combinations are derived. The sound absorption coefficient was measured using an impedance tube. As shown in Fig. 3, as the number of impregnations increased, the weight of impregnated GO increased linearly. On the other hand, the flow resistivity shows a tendency to increase rapidly as the amount of impregnation increases.

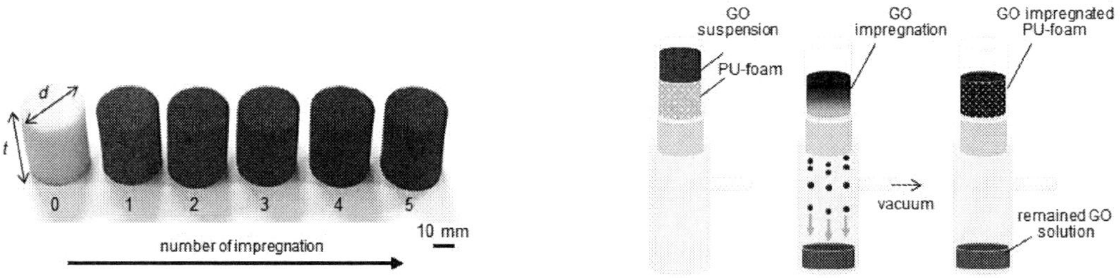

Figure 2: GO impregnated PU foam (left), illustration of vacuum impregnation (right)

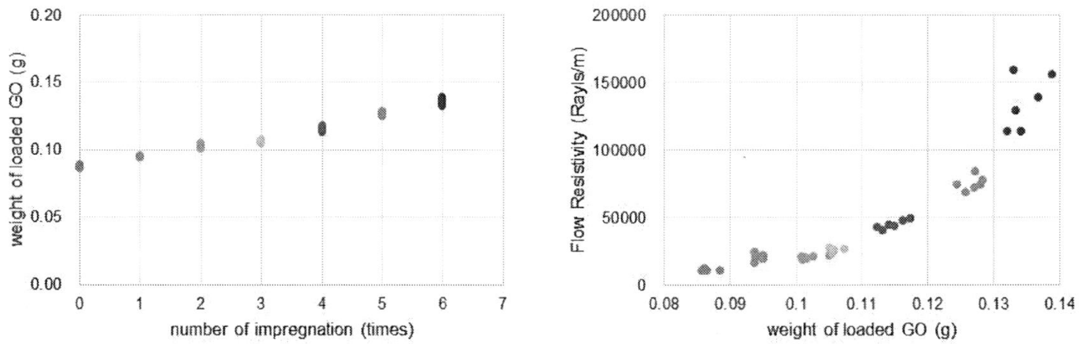

Figure 3: Weight of GO upon impregnation (left), flow resistivity as a function of loaded GO (right)

2.3 Foam properties of GO loaded PU foam

Determination of foam properties is very important in order to apply the elastic porous material theory. In this study, the bulk density, porosity, Young's modulus, and flow resistivity were measured or estimated from experiment. The tortuosity, viscous characteristic length, and thermal characteristic length were derived through fitting procedure. In the left panel of Fig. 4 shows the results of the fitting process to estimate the form properties. As shown in the right panel of Fig. 4, the fitted tortuosities are proportional to the density of impregnated GO. Table 1 below summarizes the foam properties measured or estimated according to the impregnation steps of all steps of GO impregnation. It is clearly shown in the table that Young's modulus, flow resistivity, and tortuosity increse as the number of GO impregnation increases.

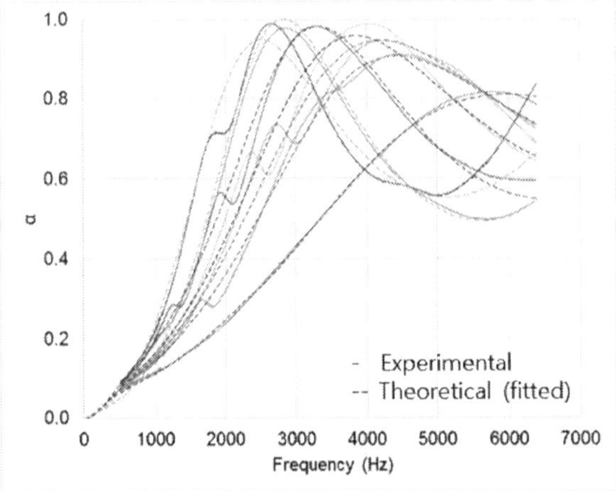

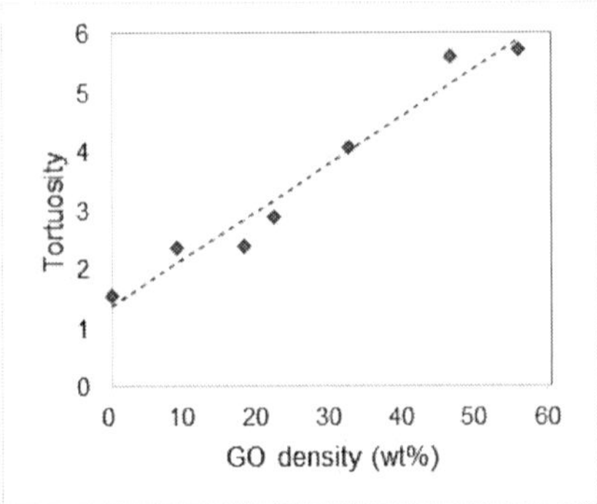

Figure 4: Fiited curves for foam parameters estimation (left), fitted tortuosites as a function of GO loading density (right)

Number of impregnation (times)	Bulk density (kg/m^3)	porosity	Young's Modulus (kPa)	Flow resistivity ($Pa\ s/m^2$)	Tortuosity	Viscous Characteristic Length	Thermal Characteristic Length
0	32.78	0.9697	48.6	10817.4	1.55	1.00E-04	2.70E-04
1	35.70	0.9682	438.7	17489.2	2.36	1.00E-04	2.00E-04
2	38.65	0.9668	831.8	20177.3	2.39	8.00E-05	2.00E-04
3	40.02	0.9661	1014.8	26039.5	2.87	9.00E-05	2.00E-04
4	43.42	0.9644	1468.8	44500.3	4.06	1.20E-04	2.00E-04
5	48.00	0.9622	2080.4	74944.4	5.59	1.00E-04	7.20E-04
6	50.99	0.9607	2479.7	130080.6	5.71	6.00E-05	6.50E-04

Table 1: measured and predicted foam properties

3 Result and Discussion

3.1 Sound absorption curves by thickness increase

Before examining the optimal combination of layers, the calculation result of the sound absorption curve with the increase in thickness was compared with the measurement result to examine the validity of the theory and measurement. Each layer was set as 4mm thick 5 times of GO impregnated PU foam. The sound absorption curves were measured from 1 layer up to 6 layer (24 mm) and compared with the calculation. As shown in the following results, it can be seen that the experimental results are well explained even by simply inputting foam properties without applying a separate optimization. Therefore, we can be sure that the subsequent optimization process can have sufficient probability with the experiment.

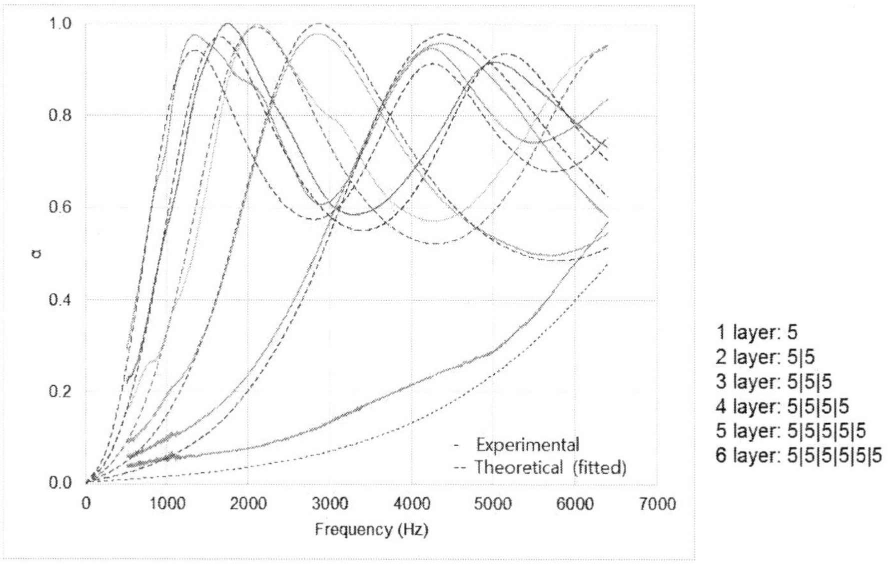

Figure 5: 1-6 layer (1 layer thickness 4mm) multi-layer sound absorbing material prediction and experimental results

3.2 Optimizing sound absorbing performance of two-layer arrangement

First, optimization was performed on the two-layer structure. In the upper graph of Fig. 6, 3 groups of curves correspond to the case of no impregnated case, 3 times GO impregnated case, and 6 times GO impregnated case, respectively. (from bottom to top). As a result of calculating the improvement of the average sound absorption rate, compared to the case without impregnation, 6 times of GO impregnated case exhibited 153% improvement of $\overline{\alpha}$ by calculation , and 148% by experiment.

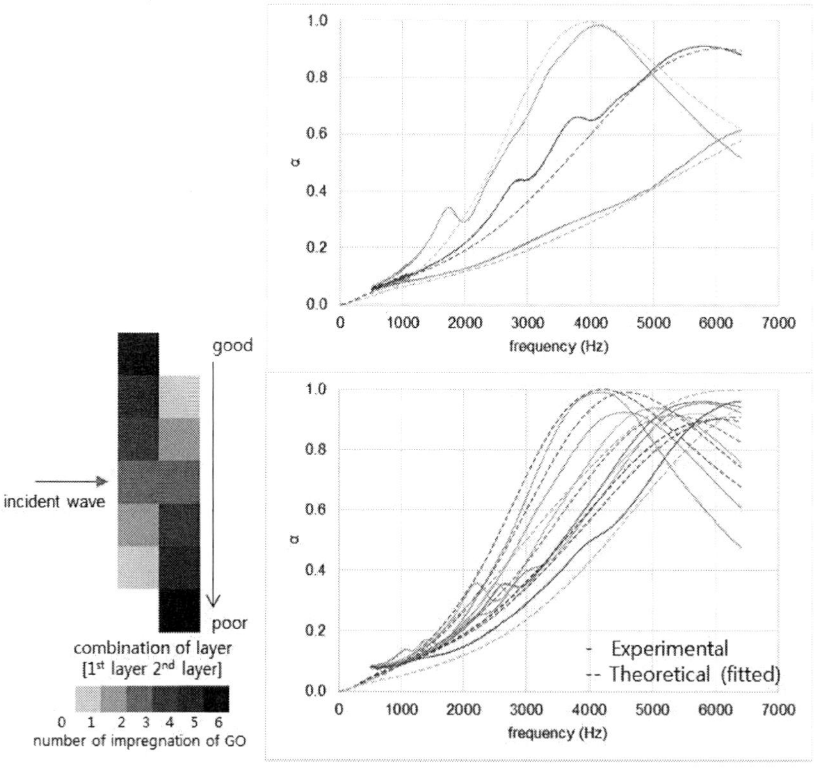

Figure 6: Optimizing sound absorption coefficient by two-layer arrangement

In the lower panel of Fig. 6, the result of fixing the total impregnation amount and optimizing the sound absorption coefficient is shown. In this case, it can be seen that it is advantageous to arrange the sound absorbing material having a high GO impregnation amount in the direction in which sound waves are incident. As a result of examining the improvement of the average sound absorption rate compared to the case of uniformly impregnated, 6|0 combination (6 times impregnation | no impregnation) exhibited 49% improvement of $\overline{\alpha}$ by calculation, and 41% by experiment.

3.3 Optimizing sound absorbing performance of multi-layer arrangement

The optimal structure of the sound absorbing material was examined while increasing the thickness. As a result, it can be seen that in the case of a thin sound absorbing material of about 10 mm, it is advantageous to arrange the sound absorbing material having a high GO impregnation amount in the direction in which sound waves are incident to improve the average sound absorption rate. In the case of a thickness of about 25 mm, it can be seen that it is advantageous to arrange the sound absorbing material having a high GO impregnation amount at two parts of the midpoint of the sound absorbing material and the incident surface. It can be seen that as the thickness approaches 50 mm, it is advantageous to place a sound absorbing material having a high GO impregnation amount on a wall where sound waves are reflected. If the average absorption coefficient derived at the maximum thickness is shown, it can be obtained that the average absorption coefficient is close to 0.9 in a sound absorbing material of about 50 mm.

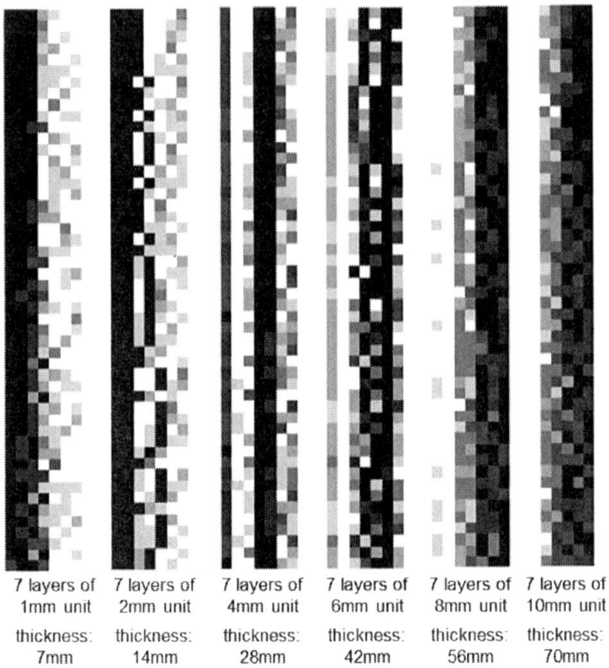

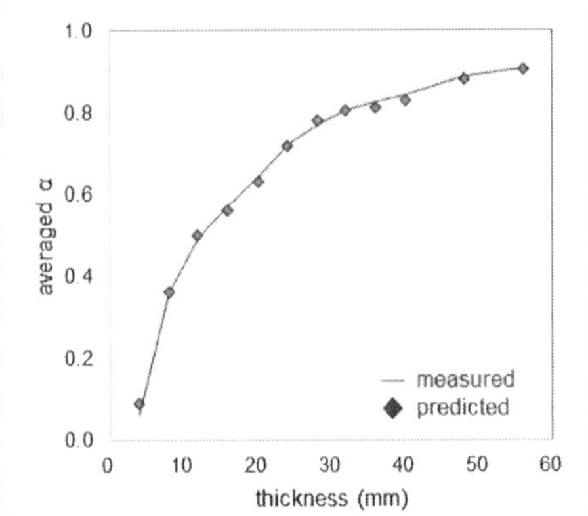

Figure 7: Optimum combinations of layer for maximizing avreage sound absoring coefficient at different thickness (left), maximized average sound absorption coefficient prediction and measurement result (right)

3.4 Summary of improvement of average sound absorption coefficient

The following is a result showing the improvement of the sound absorption coefficient according to the optimal arrangement of the sound absorbing material for each thickness. As the thickness of the layer increases, it can be seen that the improvement amount tends to decrease. In particular, it can be seen that even when the impregnation amount is fixed, the average sound absorption coefficient can be enhanced through an appropriate layer combination.

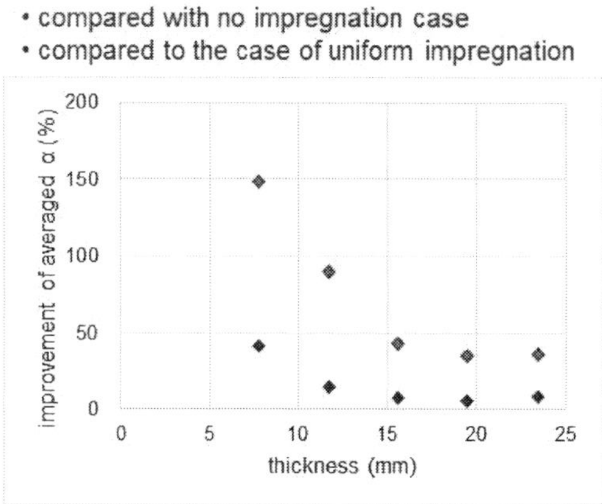

Figure 8: Improvement in average sound absorption coefficient (compared with no impregnation case, compared to the case of uniform impregnation)

3.5 Result on maximizing average sound absorption coefficient

The following shows the sound absorption coefficient curve when the maximum average sound absorption coefficient is derived. The case showing the maximum sound absorption coefficient was shown in combination 0|0|2|2|5|6|6. Here, the thickness of each layer is 8 mm. In addition, similar levels of sound absorption coefficients were also obtained in combinations with similar patterns. The theoretical average sound absorption coefficient calculation result was 0.9074, and the experimental result was 0.9011. In the combination of the same impregnation amount, the combination with the lowest average sound absorption rate was represented by 6|6|6|0|0|3|0, and similar sound absorption characteristics were also found in combinations showing similar patterns. In this case, the average sound absorption coefficient was calculated as 0.6529 and was measured to be 0.7277.

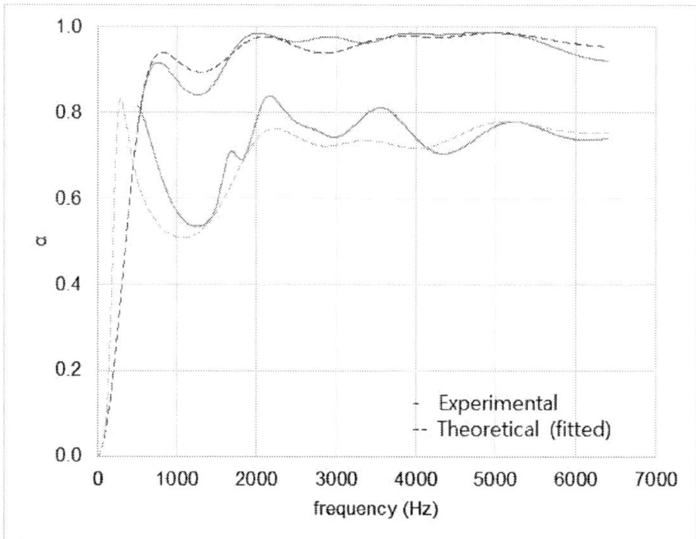

Figure 9: Maximized calculated and measured sound absorption coefficient curves of 7 layers combination (each layer is 8mm, total thickness is 56mm.)

4 Conclusion

A multi-layered layer model is proposed to optimize the sound absorption rate of the sound absorbing material. As a method of maximizing the sound absorption rate by adjusting the GO impregnation amount of each layer is proposed. The procedure for calculating the sound absorption rate using the matrix'calculation method is developed based on the theory proposed by Bolton J, Shiau N-M, and Kang Y. Predicted values of sound absorption coefficient are very consistent with the experimental values. As a result of the theoretical calculation of the improvement rate of the sound absorption rate, an improvement of 153% was expected compared to the case without impregnation, and it was measured to be 148%. For uniform impregnation, a 49% improvement was predicted and a 41% improvement was measured. The optimum arrangement of the 56 mm thick sound absorbing material was calculated as 0|0|2|2|5|6|6 and similar patterns. The average sound absorption coefficient was calculated as 0.9074 and measured as 0.9011.

5 Acknowledgements

We acknowledge the financial support from the National Research Foundation of Korea (NRF-2020R1F1A1060410).

References

[1] JP Arenas, MJ Crocker, "Recent trends in porous sound-absorbing materials" Sound & Vibration, 2010.

[2] X. Sagartzazu, L. Hervella-Nieto, J. M. Pagalday, "Review in sound absorbing materials" Archives of Computational Methods in Engineering, 2008.

[3] U. Berardi, G. Iannace, "Acoustic characterization of natural fibers for sound absorption applications", Building and Environment, 2015.

[4] Xinzhao, L. Guoming, L. Dongyan, S. Guoxin, Y. Rui, "Electrically conductive graphene-coated polyurethane foam and its epoxy composites", Composites Communications, 2018.

[5] Chen, S., Jiang, Y., Chen, J., and Wang, D, "The Effects of Various Additive Components on the Sound Absorption Performances of Polyurethane Foams", Advances in Materials Science and Engineering, 1–9, 2015.

[6] Ju Hyuk Park, Sei Hyun Yang, Hyeong Rae Lee, Cheng Bin Yu, Seong Yeol Pak, Chi Sung Oh, Yeon June Kang, Jae Ryoun Youn, "Optimization of low frequency sound absorption by cell size control and multiscale poroacoustics modeling", Journal of Sound and Vibration, 397, 17–30, 2017.

[7] Choe H, Lee JH, Kim JH, "Polyurethane composite foams including $CaCO_3$ fillers for enhanced sound absorption and compression properties", Composites Science and Technology, 194, 7, 2020.

[8] Nine, M. J., Ayub, M., Zander, A. C., Tran, D. N. H., Cazzolato, B. S., and Losic, D, "Graphene Oxide-Based Lamella Network for Enhanced Sound Absorption", Advanced Functional Materials, 2017.

[9] Jaehyuk Lee, Inhwa Jung, "Tuning sound absorbing properties of open cell polyurethane foam by impregnating graphene oxide", Applied Acoustics, 151, 10-21, 2019.

[10] Carolina Simón-Herrero, Nieves Peco, Amaya Romero, José L. Valverde, Luz Sánchez-Silva, "PVA/nanoclay/graphene oxide aerogels with enhanced sound absorption properties", Applied Acoustics, 2020.

[11] Liu, L., Chen, Y., Liu, H., Ur Rehman, H., Chen, C., Kang, H., and Li, H, "Graphene Oxide and Functionalized Carbon Nanotubes Based Semi-Open Cellular Network for Sound Absorption", Soft Matter, 2019.

[12] Wu, C. M., and Chou, M. H, "Sound absorption of electrospun polyvinylidene fluoride/graphene membranes" European Polymer Journal, 2016.

[13] Qamoshi, K., and Rasuli, R, "Subwavelength structure for sound absorption from graphene oxide-doped polyvinylpyrrolidone nanofibers", Applied Physics A, 2016.

[14] Oh, J.-H., Lee, H. R., Umrao, S., Kang, Y. J., and Oh, I.-K, "Self-aligned and hierarchically porous graphene-polyurethane foams for acoustic wave absorption", Carbon, 147, 510–5189, 2019.

[15] Bolton J., Shiau N-M., Kang Y. J. Sound transmission through multi-panel structures lined with elastic porous materials. J Sound Vib 191 (3):317–47, 1996.

[16] Kang Y. J., Bolton J. S. Finite element modeling of isotropic elastic porous materials coupled with acoustical finite elements. J Acoust Soc Am 98 (1):635–43, 1995.

2D Wrinkled Graphene Oxide-Wrapped 3D Auxetic foam for Sound Wave and Shock Energy Multi-Dissipation

J.-S. Kim[1], J.-H. Oh[2*], and I.-K. Oh[1*]

[1]National Creative Research Initiative Center for Functionally Antagonistic Nano-Engineering, Department of Mechanical Engineering, School of Mechanical and Aerospace Engineering, Korea Advanced Institute of Science and Technology (KAIST), Daehak-ro 291, Yuseong-gu, Daejeon, 34141, Republic of Korea
* Corresponding author, E-mail address: ikoh@kaist.ac.kr (I.-K. Oh)

[2]Department of Mechanical Engineering, Wonkwang University, 460 Iksan-daero, Iksan-si, Jeollabuk-do, 54538, Republic of Korea
* Corresponding author, E-mail address: junghwan5@wku.ac.kr (J.-H. Oh)

1 Abstract

Recently, noise pollution has been a serious environmental problem worldwide because the road network is expanded and the number of vehicles is rapidly increasing due to the rapid industrialization and the development of transportation such as automobiles, trains, and buses. Noise pollution can be considered serious especially in densely populated urban environments, which can significantly reduce work efficiency in research areas and schools. There is also the privacy issue in residential areas such as apartments where many households coexist. It is also known to cause disorders such as hearing loss and mental instability.

Therefore, the development of cellular structures has attracted increasing attention in practical engineering applications such as a lightweight sound absorbing panel and shock energy dissipation for preventing noise pollution. Theoretical analysis and experimental verification of sound transmission and acoustic properties through the porous media have been dedicatedly studied.[1-4] Even though cellular structures has gained attention due to their controlled physical and mechanical properties, conventional cellular structures for sound and shock energy absorbers cannot meet demands for high-energy absorbing capacity because of the limited structural properties.[5]

Research on composites foam has been introduced to improve sound absorption performance of porous foam by inserting filler materials. For examples, the enhancement of sound absorption in the entire frequency range was confirmed by inserting natural filler materials such as the tea-leaf fiber, luffa cylindrical, bamboo leaves particle, and rice hull into polymeric foam.[6-8] As interest in carbon materials has increased greatly, many studies by using carbon materials such as carbon black,[9] carbon nanotube,[10] and graphene oxide[11] as filler materials for functionality and enhancement of sound absorbing performance have been widely conducted.

Furthermore, auxetic foam with a counter-intuitive behavior has additional beneficial properties including an indentation resistance [12-14] and enhanced acoustic property [15-18], making it suitable for sound insulation applications. The difference of wave dispersion properties between conventional and auxetic foam was explained by a resonance at a lower cut-off frequency of auxetic foam because of tortuous sound wave path generated by bent struts from thermally compressed and permanently deformed framework.[19] In addition to improving the sound absorption performance in the low frequency band, auxetic porous structure has advantages including a great enhancement of shock energy dissipation and load bearing capability compared to the conventional materials.

Herein, we have suggested a simple, facile and effective method to fabricate two-dimensional (2D) wrinkled graphene oxide (GO)-wrapped three-dimensional (3D) auxetic foam for multi-energy dissipation materials. The suggested auxetic graphene oxide-polyurethane foam (AGPUF) can be fabricated by placing the polyurethane foam into a designated aluminium frame in a hot-pressing machine and compressing it at a temperature of 150°C for 1 hour to make the auxetic foam, and then dip-coating the auxetic foam on a graphene oxide solution with different ratios.

The proposed AGPUF has been demonstrated to show a superior sound absorbing performance of 99.7% at a frequency of 2,236 Hz and an impact resistance of 189% during a low-velocity impact test. These enhancements could be attributed to the advantage of synergistic effects between 3D auxetic foam with a negative Poisson's ratio and 2D wrinkled graphene oxide as a filler material. The auxetic hierarchically polyurethane foam has a convoluted structure with a very high tortuosity, which not only greatly improves the load-bearing capacity, but also has the effect of shifting the peak value of the sound absorption coefficient to a lower frequency and increasing the value. The 2D corrugated graphene oxide structures in cellular structures connect between the open-cell frameworks and induce the semi-open cell porous structure with an increase of the tortuosity in the porous media, resulting in a good sound absorbing performance due to the frictional energy dissipation. The advantages of AGPUF's compressibility, light weight, cost-effectiveness, and structural stability are able to open up the possibility of rapid mass production and industrialization by the rational design of heterostructures, enabling the realization of nano-micro multi-scale cellular structures. We believe that it will be promoted in the fields of sound, vibration and shock energy dissipation.

References

[1] Bolton JS, Shiau NM, Kang YJ. SOUND TRANSMISSION THROUGH MULTI-PANEL STRUCTURES LINED WITH ELASTIC POROUS MATERIALS. Journal of Sound and Vibration. 1996;191(3):317-47.

[2] Perrot C, Chevillotte F, Tan Hoang M, Bonnet G, Bécot F-X, Gautron L, et al. Microstructure, transport, and acoustic properties of open-cell foam samples: Experiments and three-dimensional numerical simulations. Journal of Applied Physics. 2012;111(1):014911.

[3] Song BH, Bolton JS. A transfer-matrix approach for estimating the characteristic impedance and wave numbers of limp and rigid porous materials. The Journal of the Acoustical Society of America. 2000;107(3):1131-52.

[4] Venegas R, Umnova O. Acoustical properties of double porosity granular materials. The Journal of the Acoustical Society of America. 2011;130(5):2765-76.

[5] Zhu H, Fan T, Peng Q, Zhang D. Giant Thermal Expansion in 2D and 3D Cellular Materials. Advanced Materials. 2018;30(18):1705048.

[6] Wang Y, Zhang C, Ren L, Ichchou M, Galland M-A, Bareille O. Influences of rice hull in polyurethane foam on its sound absorption characteristics. Polymer Composites. 2013;34(11):1847-55.

[7] Chen S, Jiang Y. The acoustic property study of polyurethane foam with addition of bamboo leaves particles. Polymer Composites. 2018;39(4):1370-81.

[8] Bülent Ekici AK, Haluk Küçük. Improving Sound Absorption Property of Polyurethane Foams by Adding Tea-Leaf Fibers. Archives of acoustics. 2012;37(4):515-20.

[9] Zhang S, Zhang G, Qiu J, Jiang Z, Xing H, Li M, et al. The effect of nanosized carbon black on the morphology and sc-CO2 foaming behavior of LLDPE/PS blends at semi-solid state. Composites Communications. 2018;7:30-5.

[10] Bandarian M, Shojaei A, Rashidi AM. Thermal, mechanical and acoustic damping properties of flexible open-cell polyurethane/multi-walled carbon nanotube foams: effect of surface functionality of nanotubes. Polymer International. 2011;60(3):475-82.

[11] Oh J-H, Lee HR, Umrao S, Kang YJ, Oh I-K. Self-aligned and hierarchically porous graphene-polyurethane foams for acoustic wave absorption. Carbon. 2019;147:510-8.

[12] Evans KE, Alderson A. Auxetic Materials: Functional Materials and Structures from Lateral Thinking! Advanced Materials. 2000;12(9):617-28.

[13] Evans KE, Alderson KL. Auxetic materials: the positive side of being negative. Engineering Science and Education Journal. 2000;9(4):148-54.

[14] Lakes RS, Elms K. Indentability of Conventional and Negative Poisson's Ratio Foams. Journal of Composite Materials. 1993;27(12):1193-202.

[15] Scarpa F, Ciffo LG, Yates JR. Dynamic properties of high structural integrity auxetic open cell foam. Smart Materials and Structures. 2003;13(1):49-56.

[16] Scarpa F, Smith FC. Passive and MR Fluid-coated Auxetic PU Foam – Mechanical, Acoustic, and Electromagnetic Properties. Journal of Intelligent Material Systems and Structures. 2004;15(12):973-9.

[17] Scarpa F, Bullough WA, Lumley P. Trends in acoustic properties of iron particle seeded auxetic polyurethane foam. Proceedings of the Institution of Mechanical Engineers, Part C: Journal of Mechanical Engineering Science. 2004;218(2):241-4.

[18] Wilde WP, Brebbia CA, Hernández S, Technology WI, Brussel VU, da Coruña U. High Performance Structures and Materials VI: WIT Press; 2012.

[19] Chen CP, Lakes RS. Micromechanical Analysis of Dynamic Behavior of Conventional and Negative Poisson's Ratio Foams. Journal of Engineering Materials and Technology. 1996;118(3):285-8.

Impedance-tube characterisation of additively manufactured slitted sound absorbers

K. C. Opiela[1], T. G. Zieliński[1], K. Attenborough[2]

[1] Institute of Fundamental Technological Research, Polish Academy of Sciences, Pawińskiego 5B, 02-106 Warsaw, Poland
[2] The Open University, School of Engineering and Innovation, Milton Keynes MK7 6AA, UK

An acoustical characterisation of additively manufactured rigid slitted structures is considered. A set of six JCAL microstructural parameters is deduced from dynamic density and bulk modulus obtained from normal incidence surface acoustic impedance experimental data. The results show that the characteristic lengths are the most difficult to characterise.

1 Introduction

Sufficiently dense simple slitted structures have a tendency to absorb sound as efficiently as more sophisticated porous materials [1, 2]. The normal incidence sound absorption coefficient pertinent to several additively manufactured (i.e. 3D printed) slitted samples was studied in [3]. Rather poor agreement between predictions and measurement data was observed when slits were perforated in the direction perpendicular to the sample axis. To identify possible sources of discrepancies, experimental complex density, $\varrho_{\rm eq}(\omega)$, and bulk modulus, $K_{\rm eq}(\omega)$, are calculated and investigated ($\omega \equiv 2\pi f$, f is the temporal frequency). An acoustical characterisation of *six* Johnson-Champoux-Allard-Lafarge (JCAL) parameters that is based on $\varrho_{\rm eq}(\omega)$ and $K_{\rm eq}(\omega)$ is performed. The estimated parameters are: the open porosity, ϕ, (static) viscous permeability, k_0, (static) thermal permeability, k'_0, (inertial) tortuosity, α_∞, and two characteristic lengths: viscous, Λ, and thermal, Λ'.

2 Slitted geometries

Two microstructures with parallel straight slits of width 0.3 mm normal to the incident surface and separated from each other by 0.4-mm thick solid strips were designed and 3D printed: a configuration with and without a periodic cylindrical perforation of diameter 2 mm and period 4.95 mm, perpendicular to the slit plane (see pictures in Figures 1b and 1c). The non-perforated sample was manufactured in the Fused Deposition Modelling technology, whereas the perforated one was produced from a cured resin in the UV LCD technology. Both samples had diameter of 29 mm and height of 49.5 mm.

3 Results of acoustical characterisation

A procedure to evaluate complex density and bulk modulus functions based on impedance-tube measurements, known as the two-cavity method, was proposed in 1989 [4]. According to it, $\varrho_{\rm eq}(\omega)$ and $K_{\rm eq}(\omega)$ are uniquely related to the normal incidence surface acoustic impedances obtained for a sample with two distinct air cavities behind. Knowing the sample height (i.e. material thickness) and some air properties, one computes $\varrho_{\rm eq}(\omega)$ and $K_{\rm eq}(\omega)$ that serve as an input for acoustical characterisation. The set of six geometrical parameters governing visco-inertial as well as thermal dissipation of sound waves in porous media described by the JCAL model is determined following References [5, 6, 7].
Table 1 shows the JCAL parameters characterised using the method described above (with air cavities of 20 mm and 40 mm, and 110 mm and 130 mm for the slitted sample with and without perforation, respectively), and computed on idealised cells representative for the two studied microgeometries by performing Stokes, Laplace and Poisson finite element analyses (see [3]). The corresponding normalised complex density and bulk modulus functions are plotted in Figure 1.
In both cases, there is a qualitative agreement between the numerically predicted, characterised, and experimental $\varrho_{\rm eq}(\omega)$ and $K_{\rm eq}(\omega)$. However, the discrepancy in the obtained model parameters, especially in Λ and Λ', suggests the difficulty of the approach in precise determination of true geometrical features of additively manufactured porous materials.

Acknowledgements

The financial support of the project number 2020/37/N/ST8/04071: "Impact of the 3D printing process on the acoustic properties of porous materials", financed by the National Science Centre, Poland, is gratefully acknowledged.

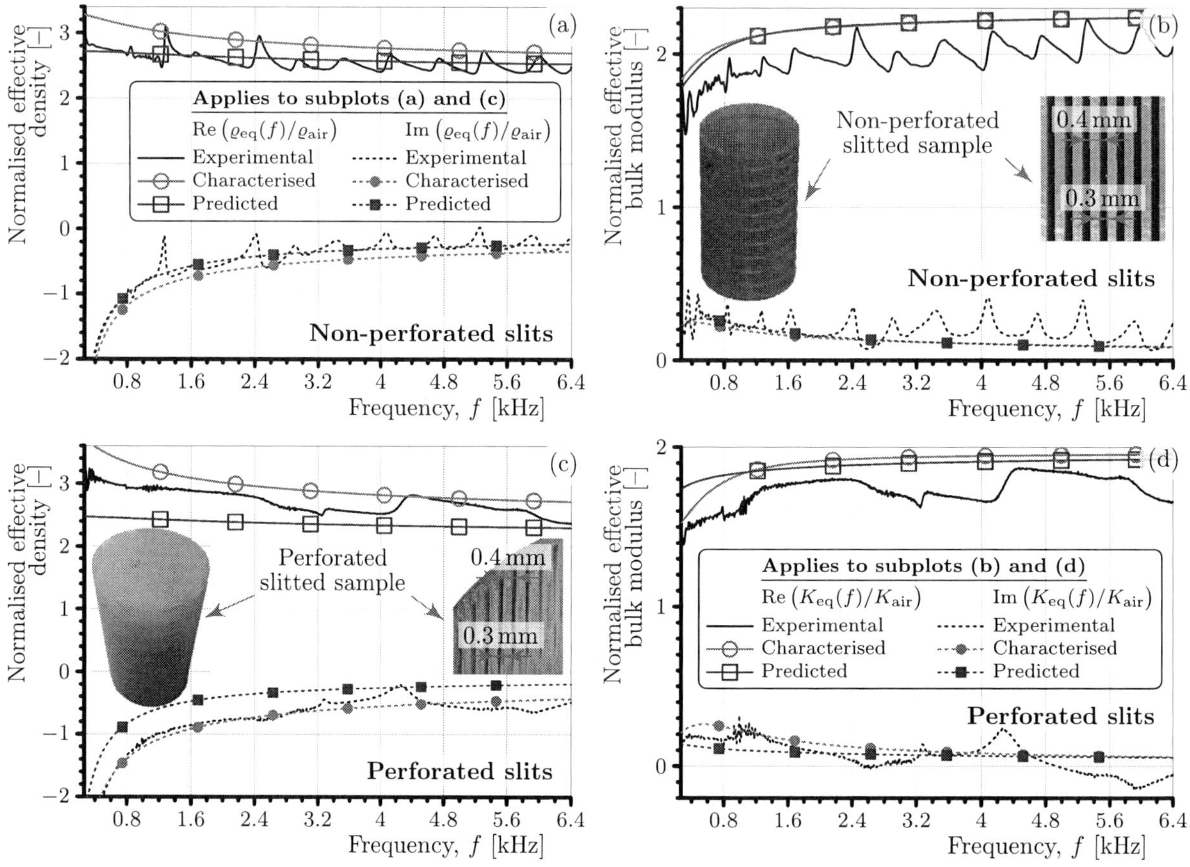

Figure 1: Normalised effective mass density (a,c) and bulk modulus (b,d) for the slitted structure without (a,b) and with (c,d) the cylindrical perforation. In the legends, ϱ_{air} is the mass density of air, and K_{air} denotes the bulk modulus of air.

	Geometrical parameters					
Geometry, case	ϕ [–]	k_0 [$10^{-9}\,\mathrm{m}^2$]	k_0' [$10^{-9}\,\mathrm{m}^2$]	α_∞ [–]	Λ [mm]	Λ' [mm]
Non-perforated, computed	0.4286	3.214	3.214	1.000	0.3000	0.3000
Non-perforated, characterised	0.4250	3.350	3.800	1.010	0.2000	0.2600
Perforated, computed	0.5018	3.915	20.65	1.070	0.3218	0.3806
Perforated, characterised	0.5000	3.311	4.000	1.150	0.1500	0.5000

Table 1: Computed and characterised geometrical model parameters.

References

[1] K. Attenborough, *Microstructures for lowering the quarter wavelength resonance frequency of a hard-backed rigid-porous layer*, Appl. Acoust., 130 (2018), pp. 188–194.

[2] K. Attenborough, *Macro- and micro-structure designs for porous sound absorbers*, Appl. Acoust., 145 (2019), pp. 349–357.

[3] K. C. Opiela, T. G. Zieliński, and K. Attenborough, *Manufacturing, modeling, and experimental verification of slitted sound absorbers*, in Proc. of ISMA2020 International Conference on Noise and Vibration Engineering/USD2020 International Conference on Uncertainty in Structural Dynamics, W. Desmet, B. Pluymers, D. Moens, and S. Vandemaele, eds., KU Leuven, Belgium, 2020, pp. 409–420.

[4] H. Utsuno, T. Tanaka, T. Fujikawa, and A. F. Seybert, *Transfer function method for measuring characteristic impedance and propagation constant of porous materials*, J. Acoust. Soc. Am., 86 (1989), pp. 637–643.

[5] L. Jaouen, E. Gourdon, and P. Glé, *Estimation of all six parameters of Johnson-Champoux-Allard-Lafarge model for acoustical porous materials from impedance tube measurements*, J. Acoust. Soc. Am., 148 (2020), pp. 1998–2005.

[6] X. Olny and R. Panneton, *Acoustical determination of the parameters governing thermal dissipation in porous media*, J. Acoust. Soc. Am., 123 (2008), pp. 814–824.

[7] R. Panneton and X. Olny, *Acoustical determination of the parameters governing viscous dissipation in porous media*, J. Acoust. Soc. Am., 119 (2006), pp. 2027–2040.

Parameter study on a composite sound-absorbing structure lined on elevator shafts

Ting Qu[1], Hequn Min[2*]

(1.School of Architecture, Southeast University, Nanjing 210096, China;)

(2.Key Laboratory of Urban and Architectural Heritage Conservation of Ministry of Education, Southeast University, Nanjing 210096, China)

Abstract: In order to reduce the noise in elevator shafts, a composite sound-absorbing structure which can be applied to inner walls of elevator shafts is proposed. Based on the theory of acoustic impedance of porous material proposed by K. Attenborough and the theory of acoustic impedance of perforated plates proposed by Bies, parameter optimization of the sound-absorption structure is discussed in the paper. The composite structure, as shown in Figure 1, is composed of damping gypsum mortar board, composite rock wool board and cement mortar perforated board . Based on theoretical analysis, numerical simulation and experimental measurement, parameter optimization of the structure is discussed on the structure normal incidence sound absorption performance, based on the influence of parameters, such as thickness, flow resistance and porosity of the rock wool board, and perforation rate, thickness and density of the perforated board. Results show that porosity, flow resistance and bulk density of the rock wool board, and the perforation rate and thickness of the perforated board play a key role in the structure sound-absorbing performance. Increasing the thickness of the perforated board, reducing the perforation rate and increasing the porosity of the porous material can enhance absorption performance at low- frequencies, as shown in Fig. 2-3. Normal incidence sound absorption coefficient of the optimized composite structure may reach 0.6 and 0.8 600Hz and 1600Hz, respectively. It is indicated that, there is great application potential for the proposed composite sound-absorbing structure has in elevator shaft noise control.

Keywords: Elevator shaft; Sound absorbing structure; Porous material; Flow resistance; Acoustic impedance

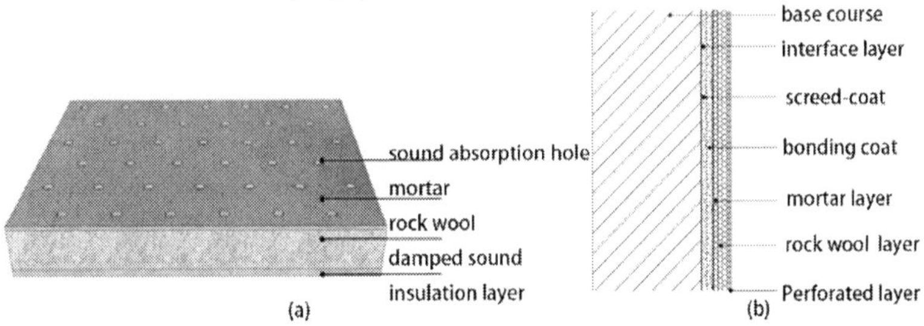

Fig.1 (a) A diagram of the composite sound-absorbing structure; (b) The application of the composite sound-absorbing structure lined on elevator shafts

* Corresponding author.
E-mail address: hqmin@seu.edu.cn

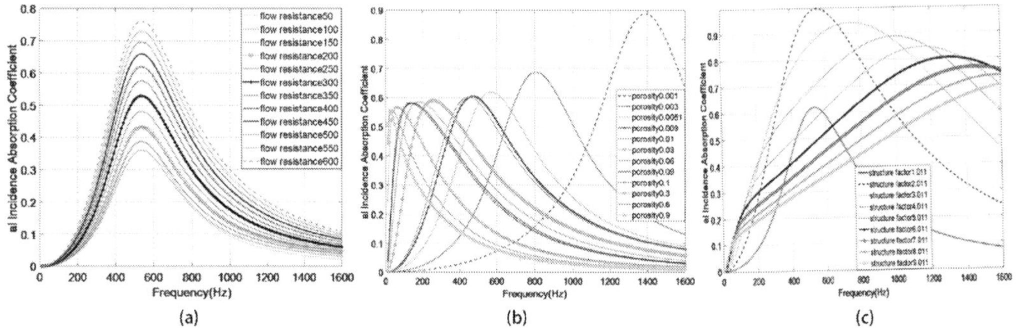

Fig.2 The effect of (a) flow resistance, (b) porous rates, (c) structure factor of porous material on sound absorption coefficient of composite structure

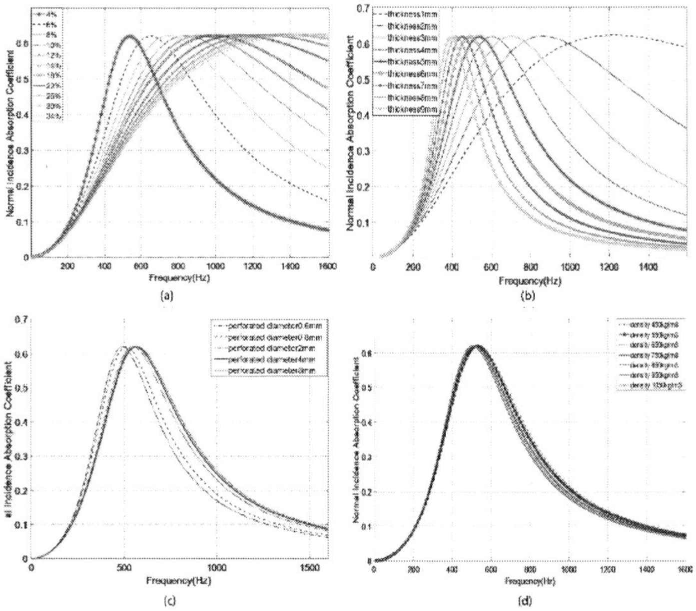

Fig.3 The effect of (a) perforated rates, (b) thickness, (c) perforated diameter, (d) density of perforated panel on sound absorption coefficient of composite structure

参考文献

[1] K. Attenborough. Ground parameter information for propagation modeling. Journal of Acoustical Society of America, 1992, 92: 418–427.

[2] David A. Bies, Colin H. Hansen Engineering noise control theory and practice. London and New York: spon press taylor francls group 2013, 636-637.

[3] Cui Y Q. Study on the preparation and modeling process of foam-sandwich structure acoustic composites. Harbin Engineering University, 2019

[4] Ma D Y. Engineering manual for noise and Vibration control. Beijing: Machinery Industry Press, 2002

[5] ISO 10534-2:1998. Acoustics — Determination of sound absorption coefficient and impedance in impedance tubes — Part 2: Transfer-function method

Acoustic Design and Optimization of Automotive Inner Dash Insulations Using Compound Injection Moulding (CIM) Composites.

[1]G. Lesage, M. Tarello, F. Tinti. R. Morris-Kirby, J. Zepponi, E. Harry

[1]Adler Pelzer Group

1 Introduction

The inner dash insulation is a fundamental component of an automotive acoustic refinement package. Acoustic performance depends, not only, on the material combinations used but also the accuracy of fitment and finish in conjunction with the associated pass-through and ancillaries. As OEM's strive to progressively reduce noise from power units and drive trains it has become necessary to introduce enhanced levels of detail into the design and manufacturing of the inner dash insulation. A popular design uses the concept of a "mass – spring" composite. The mass consists of an impervious high-density layer with the spring made from a porous-elastic spacing material typically using fibres or Polyurethane foam. An extension of this technique used by APG, called Compound Injection Moulding (CIM), allows for the localized placement of mass instead of a uniform distribution.

This paper compares a traditional manufacturing solution with one manufactured using CIM.

2 Acoustic Concepts

The inner dash insulation is located behind the dashboard, or instrument panel, and sits on the cabin side of the bulkhead (firewall) between the cabin and engine bay. It must accommodate the multitude of attached components, such as HVAC and ECU as well as the steering column and operating pedals. Although the inner dash is hidden from the driver and passengers, the ultimate refinement of the vehicle depends on it. Whilst the body-in-white of the vehicle provides an inherent noise transmission reduction depending on the thickness of the construction material (steel, aluminium or plastic composites) and the mass of any additional structural features it is generally too low to provide acceptable noise reduction from high powered drive units. This is irrespective of whether an ICE or Electric drive motor is examined. Introducing an inner dash insulation can significantly enhance transmission loss especially when efficiently optimised. There are different technologies available that are based on different acoustic principles and related weight in accordance with an OEM's specifications along with production volumes and associated costs.

A popular, high performance, design uses the concept of a "mass – spring" composite. The mass consisting of an impervious high-density layer with the spring made from a porous-elastic spacing material typically using fibres or Polyurethane foam. Vacuum moulding of the mass layer, sometimes called "heavy layer" is a traditional manufacturing method. Compounding Injection moulding (CIM) technology allows the effective mass to be adjusted across the surface of the inner dash which enables fine tuning acoustic optimisation, either by testing or simulation. This freedom in design also allows the incorporation of built-in grommet sealing of pass-throughs whilst significantly reducing production scrap rates. During optimisation, the mass value of the CIM layer can also be tuned in combination with a ribbed PU foam "spring". Figure 1 compares surface velocity maps of a vacuum moulded inner dash (left) with an optimised CIM Inner dash (Right) clearly showing the reduction in surface velocity, bluer areas, afforded by the CIM technology. The effect of the CIM Inner dash on overall cabin sound pressure level during a wide-open throttle acceleration in 3rd gear is shown on figure 2 for the driver and front passenger of a 4-cylinder ICE saloon.

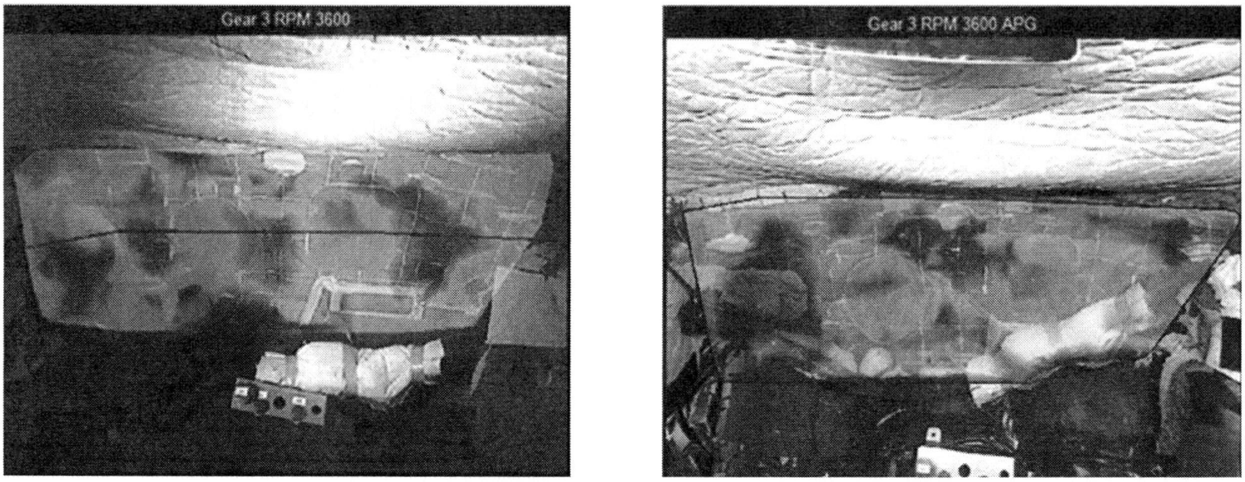

Figure 1 – Surface velocity mapping using "Scan and Paint" comparing Vacuum moulded v CIM technology – average 400 to 2kHz at 3600 rev/min. For a 4-cylinder ICE in 3rd gear.

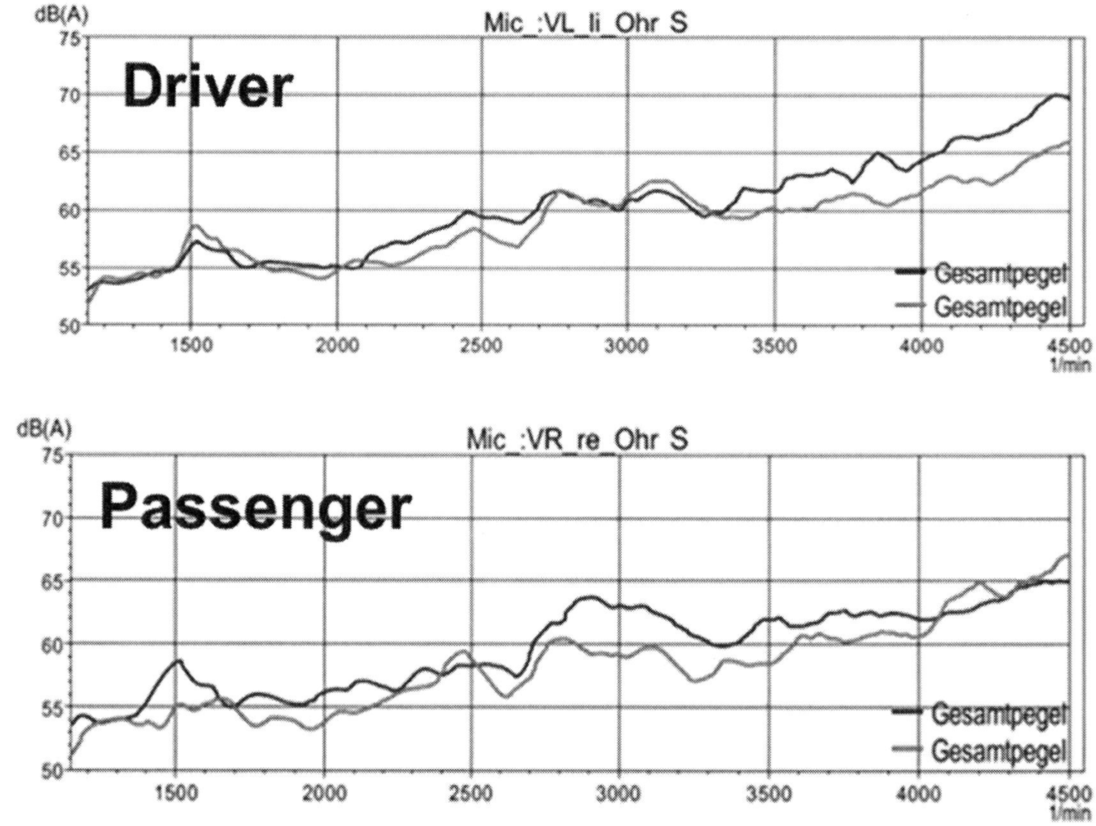

Figure 2 – Overall Sound Pressure Level dBA during 3rd gear WOT (Black Baseline – Red Optimised CIM)

3 Compound Injection Moulding Technology

Traditional vacuum moulding of the mass layer is a simplified version of thermoforming. A polymeric sheet, which is obtained from extrusion technology is heated, usually via infrared radiation, to a temperature close to its' melting temperature and then transferred into a single sided forming tool. The forming process is augmented by applying a vacuum to the underside of the tool and this is held until the material forms the appropriate final shape. Once formed the spring layer is introduced either by in situ foaming or another suitable lamination process if fibre-based materials are chosen. The complete composite is placed on a cutting fixture and perforated according to the vehicle installation requirements. Several constraints are linked to this approach as it deals with a 2-D shape which is converted into a 3-D shape and this often causes stretching of the mass layer sheet which in turn produces local undesirable density variations.

Several years ago, the concept of mass-spring dashes, using vacuum moulding, was transferred to the technically advanced Compound Injection Moulding (CIM) process. This change enabled the inherent advantages of injection moulding (undercuts, ribs etc.) whilst at the same time avoiding the cutting process of vacuum moulding and the associated material scrap rate. CIM is a combination of a typical plastic compounding process with a formulation of specific polymers, mineral fillers, dyes, stabilisers, process aids and fire retardants together with the injection moulding process. It is a one step process which not only saves significant logistic, energy and post processing costs but also enables the mass distribution to be placed according to acoustic optimisation together with associated physical and mechanical requirements.
By using different densities, it is possible to mould parts with different overall weights from the same tool. In addition to this it is possible to significantly vary all parameters of the polymeric mixture. For example, it is possible to vary the final density from 1000kg/m^3 (close to that of a non-filled mixture) to a highly filled compound of up to 2500 kg/m^3, or in case of need higher, and this flexibility considerably increases the potential acoustic performance. The elastic modulus, loss factor and other parameters can also be tailored to the requirements of the final component. Where required the part thickness can be reduced providing considerable design freedom when compared to vacuum moulding technology. In terms of environmental performance, it also allows extensive savings of virgin resources as using injection moulding does not create cutting waste whilst also offering the opportunity to recycle any scraps as the material is a thermoplastic polymer and moreover it allows substitution of adjustment/assembly parts, by integration in the CIM design, so minimizing resources due to reducing material complexity.

Hybrid Polyfoam T Ultimate: an efficient chips urethane based green & light insulator

A. Duval, M. Goret, D. Lemaire, G. Crignon

Treves Product, Services & Innovation, 2-4, rue Emile Arquès,
CS 70017 - 51686 Reims Cedex 2 - France

1 Introduction

The hybrid stiff insulators family, hard porous layer - impervious layer - soft porous layer, has been successful in the automotive industry these last years, especially for very light lay-ups. Indeed, the classically lightweight top hard layer gives mechanical stability to the insulator and is thus easy to manipulate for assembly purposes, while ensuring a mass-spring like good insulation behavior thanks to its intrinsic stiffness and to the impervious glued film-like backlayer. These "textile septum" good properties already reported present a drawback linked to their relatively low thicknesses (top hard layer) and therefore relatively bad absorption properties in the middle frequency range [1, 2]. Indeed, in order to get Young's modulus above 1 MPa typically, the usually light $1000\,g/m^2$ up to $2000\,g/m^2$ top compressed porous layers are presenting thicknesses lying between $5\,mm$ and maximum $9\,mm$. The overall available thicknesses of insulators in the automotive industry going rarely above 30 mm typically do not allow to go beyond risking then to overcompress the soft spring layer which may become too hard in that case and thus not decoupling correctly anymore!

2 Hybrid Polyfoam T Ultimate description

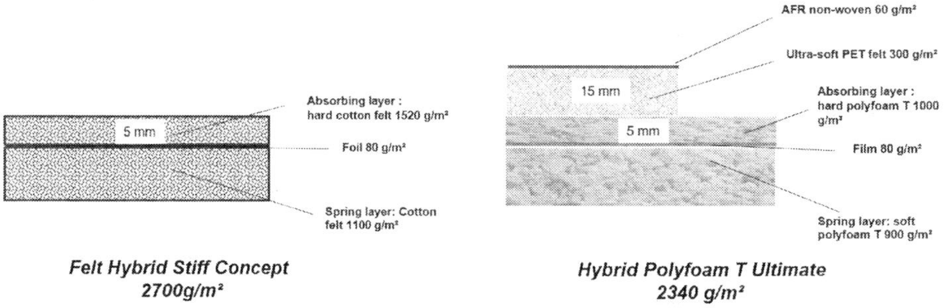

Figure 1: Hybrid Polyfoam T Ultimate typical - 15 % weight reduction

A solution consists in using an ultra-soft highly compressible PET felt out of the available thickness CAD of the insulators (with a tuned AirFlow Resistance nonwoven on top). Therefore, this ultra-soft PET felt will be locally in interference with components behind the Instrument Panel (up to 70 % or 80 % of the total PET felt thickness) like the HVAC air-conditioning unit or glove box for example etc... Figure 1 right illustrates the typical lay-up applied locally where the interferences allow it, as well as the resulting weight reduction between - 15 % to - 20 % enabled by this broadband additional absorption and Polyfoam T material (cf. Figure 2).

3 Hybrid Polyfoam T Ultimate NVH performance

Beyond the addition of this ultra-soft PET, the switch from a Felt Hybrid Stiff concept to a Hybrid Polyfoam T 80 % recycled and 100 % recyclable construction presents the advantage of drastically increasing the insulation performance for the low thicknesses for the normally soft poroelastic spring layer. Indeed, the cotton felt material becomes very hard above $200\,kg/m^3$ up to $300\,kg/m^3$ with Young's modulus between 1E6 Pa up to 1E8 Pa progressively depending on the binder percentage and nature, whereas the Polyfoam T made of chips urethane coming from shredded used mattresses remains below 5E5 Pa with an optimal percentage of Co-PET bi-component fibers (density $260\,kg/m^3$)...

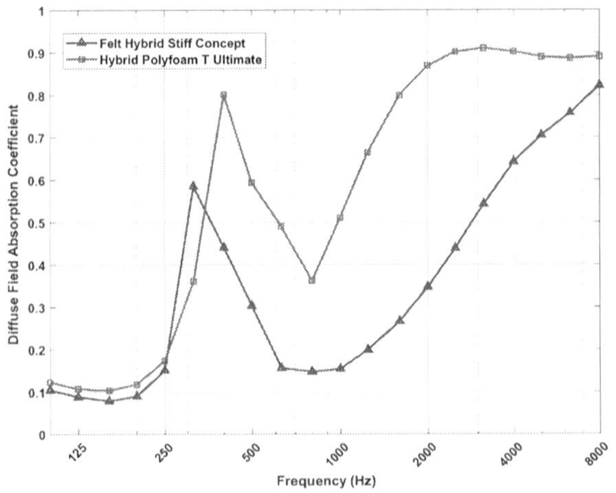

Figure 2: Hybrid Polyfoam T Ultimate Absorption Coefficient

Figure 3 shows a measured Insertion Loss improvement of 7 dB in average compared to an injected foam 58 kg/m^3, which remains relatively soft at 5 mm thickness with Young's modulus below 1E5 Pa, proving that the increased density is thus favorably contributing here. The comparison with the super hard cotton felt as poroelastic spring does not make sense (thus not reported here) and would be highly in favor of the still soft highly compressed Polyfoam T material anyway [3]...

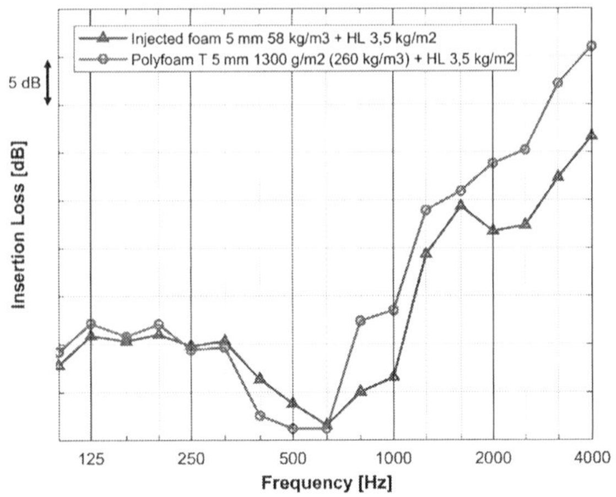

Figure 3: Polyfoam T Insertion Loss small cabin measurements at low thicknesses

4 Conclusion

Beyond its one shot competitive process, the Hybrid Polyfoam T concept (without PET felt optionally) presents the additional advantage for the hard absorbing barrier layer to be easily charged with solid inclusions in the volume up to 30 % while maintaing a good absorption (cf. [4]) or powdered below backfoiled here for even higher equivalent masses (cf. [5]), in order to increase the overall Transmission Loss and thus address upper segment vehicles. With superior middle and especially high frequency acoustic global performance within actual available spaces, the Hybrid Polyfoam T Ultimate insulator with its ultra-soft PET felt is thus particularly adapted to electric or hybrid electrified vehicles with easily tunable acoustic performances.

References

[1] A. Duval and L. Bischoff, *Stiff textiles or felts glued on light impervious layers: a new green light septum fiber technology*, SAPEM Conference, Ferrara, Italy, 2011.

[2] A. Duval, J-F. Rondeau, L. Dejaeger, F. Lhuillier and al., *Generalized Light-Weight Concepts: A New Insulator 3D Optimization Procedure*, SAE Conference - 2013-01-1947, Grand Rapids (MI), USA, 2013.

[3] J.-F. Allard and N. Atalla, *Propagation of sound in porous media: modelling sound absorbing materials*, Wiley & Sons, 2009.

[4] A. Duval, G. Crignon, M. Goret and M. Roux, *Comprehensive hybrid stiff insulators family: the chips urethane contribution*, SAE Conference - 2017-01-1883, Grand Rapids (MI), USA, 2017.

[5] A. Duval, G. Crignon, M. Goret, D. Lemaire and al., *Ecofelt Hybrid Stiff NVH Tunable Insulator*, ISNVH Conference - 2018-01-1494, Graz, Austria, 2018.

Closed-forms derivation and experimental validation of the Willis coupling of different asymmetric structures

M. Malléjac [1], A. Merkel, [2], D. Torrent [3], J. Li [4], V. Tournat [1], V. Romero-García [1], and J.-P. Groby [1]

[1] Laboratoire dAcoustique de l' Université du Mans (LAUM), UMR 6613, Institut d'Acoustique - Graduate School (IA-GS), CNRS, Le Mans Université, France
[2] Université de Lorraine, CNRS, IJL, F-54000 Nancy, France
[3] GROC, UJI, Institut de Noves Tecnologies de la Imatge (INIT), Universitat Jaume I, 12071 Castelló, Spain
[4] Department of Physics, The Hong Kong University of Science and Technology, Clear Water Bay, Hong Kong, China

We derive analytically and verify experimentally the Willis coupling parameters of simple one-dimensional asymmetric and reciprocal resonant structures made of Helmholtz resonators and plates. The coupling terms of the derived analytical closed-forms are analyzed in each of the three configurations.

1 Introduction

Bi-isotropic media [1] and their acoustic counterparts, the Willis materials [2, 4, 3], have drawn considerable interest. These materials have the particularity to couple the potential and kinetic energy in their constitutive equations, therefore enhancing the possibility of control and design for desired applications. In this talk, we will derive and analyze the closed-forms of the effective and coupling parameters for three different asymmetric and reciprocal unit cells composed of two detuned Helmholtz resonators, two plates, or one Helmholtz resonator and one plate. The closed-forms will be analyzed to investigate the influence of the nature of the asymmetry on the coupling terms.

2 Derivation of the closed-forms

One dimensional asymmetric and reciprocal media composed of a periodic arrangement of d-thick unit cells are considered. The closed-forms of the effective parameters are derived from the transfer matrix of the unit cell $\mathbf{T}$ that links the state vector $\mathbf{W} = \langle p, v \rangle^t$ (pressure p and particle velocity v) from one side to the other of the unit cell

$$\mathbf{W}(\mathbf{d}) = \mathbf{T}\mathbf{W}(0) = e^{\mathbf{A}d}\mathbf{W}(0). \tag{1}$$

The $\mathbf{A}$ matrix arises from the constitutive equations and reads for a Willis material as

$$\mathbf{A} = i\omega \begin{bmatrix} \chi & \rho \\ 1/K & -\chi \end{bmatrix}, \tag{2}$$

where ρ is the density, K the bulk modulus, and χ the Willis coupling term.
Making use of the Padé's approximation of the matrix exponential,

$$\mathbf{T} = e^{\mathbf{A}d} \approx (\mathbf{Id} - \mathbf{A}d/2)^{-1} (\mathbf{Id} + \mathbf{A}d/2), \tag{3}$$

one can approximate the matrix $\mathbf{A}$ in terms of the t_{ij} elements of the $\mathbf{T}$ matrix

$$\mathbf{A} \approx \frac{2}{d(2 + t_{11} + t_{22})} \begin{bmatrix} t_{11} - t_{22} & 2t_{12} \\ 2t_{21} & t_{22} - t_{11} \end{bmatrix}, \tag{4}$$

by ensuring the reciprocity of the transformation.
We can then identify the closed-forms of the effective parameters and Willis coupling term

$$\chi = \frac{-2i(t_{11} - t_{22})}{\omega d(2 + t_{11} + t_{22})}, \tag{5}$$

$$\rho = \frac{-4it_{12}}{\omega d(2 + t_{11} + t_{22})}, \tag{6}$$

$$[K]^{-1} = \frac{-4it_{21}}{\omega d(2 + t_{11} + t_{22})}. \tag{7}$$

3 Experimental validation of the derived closed-forms

The closed-forms of the effective parameters for the three unit cells depicted in Fig. 1 are compared against numerical simulations and experimental measurements. The measurements are performed with a 4 microphones impedance tube of radius $r = 2.5$ cm excited by a stept sine.

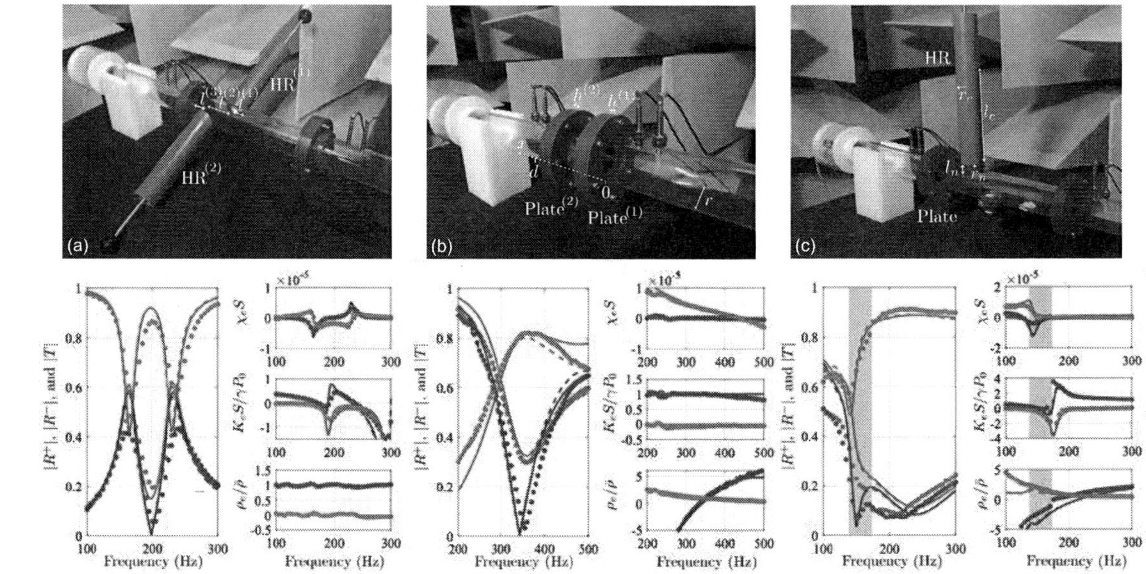

Figure 1: Photographs and results for the three cases analyzed: (a) two detuned Helmholtz resonators, (b) two detuned plates, and (c) plate and Helmoltz resonator. The different subplots show respectively the scattering magnitudes, the Willis coupling parameter, the normalized effective bulk modulus, and the normalized effective density, for each case. The solid lines represent the analytical, the dashed lines the numerical, and the symbols the experimental results.

A very good agreement is found between the three methods, thus validating the analytical derivation of the effective parameters and the Willis coupling.

4 Conclusion

This work shows a new approach for the understanding of the Willis metamaterials: the derivation of the closed-forms of the effective parameters and Willis coupling terms via the Padé's approximation of transfer matrix of different one dimensional asymmetric and reciprocal unit cells. We discuss the different types of coupling terms due to the nature of the asymmetry, either related to a detuning of identical type resonators or due to a physical asymmetry when the unit cell involves different types of resonators. A two-fold validation is performed with numerical and experimental results that matches well the derived closed-forms.

References

[1] I. Lindell, A. Sihvola, S. Tretyakov, and A. J. Viitanen. *Electromagnetic waves in chiral and bi-isotropic media*, pages 122. Artech House, 1994.

[2] J. R. Willis. *Variational principles for dynamic problems for inhomogeneous elastic media*, Wave Motion, 3:1 11, 1981.

[3] A. Merkel, V. Romero-García, J.-P. Groby, J. Li, and J. Christensen. *Unidirectional zero sonic reflection in passive PT-symmetric willis media.* Phys. Rev. B, 98:201102, 2018.

[4] M. B. Muhlestein, C. F. Sieck, P. S. Wilson, and M. R. Haberman. *Experimental evidence of Willis coupling in a one-dimensional effective material element.* Nat Commun, 8:15625, 2017.

Analytical approximations for sound absorption by labyrinthine architectures

K. Attenborough

Engineering and Innovation, The Open University,
Milton Keynes, MK7 6AA, UK

1 Introduction

Predictions and measurements have shown that labyrinthine architectures with long convoluted internal sound channels providing ‘coiled up space’ give good impedance matching [1] and yield high absorption coefficients at wavelengths much greater than the layer thickness [2]. Most computations for labyrinthine structures have used numerical methods such as FEM. But an alternative is to regard an array of narrow labyrinthine channels in a non-porous solid to be a porous material with a high tortuosity. In this paper, contiguous units each of which contains a labyrinthine slit are considered to form a rigid-porous solid acting, acoustically, as an effective fluid layer. Analytical approximations for the acoustical properties of the labyrinth make use of the classical result for sound propagation in tortuous slits [3]. Dual porosity theory is used to predict the results of making a labyrinthine perforation in a porous matrix having high flow resistivity and tortuosity. Example predictions for 4.5 cm thick, porous layers containing labyrinthine slit perforations, suggest the labyrinthine slit perforations in a high flow resistivity matrix have the potential for achieving narrow deep subwavelength absorption peaks. The presented approximations should be useful in designing labyrinthine metamaterials with targeted acoustical performance.

2 Labyrinthine slits parallel and normal to the surface

Consider a plane wave incident on a layer of contiguous rectangular units each of which contains a single labyrinthine slit of width (b) with N folds in the slit formed by walls of length l and width w. The walls are either parallel to (Figure 1(a)) or normal to (Figure 1(b)) the surface on which the wave is incident.

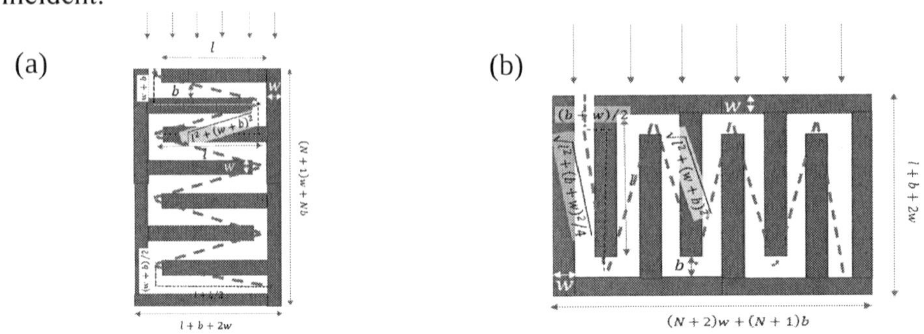

Figure 1: (a) Units containing a single labyrinthine slit, formed by walls either (a) parallel or (b) normal to the surface, showing the associated dimensions and approximate streamline paths.

The shortest path fluid streamline paths through the slit labyrinths are approximated by zig zags (red broken lines in Figures 1(a) and (b)). Expressions for tortuosity are respectively:

$$T_H=\left\{\frac{(N-1)\sqrt{(l+b)^2+(b+w)^2}+\sqrt{(l+b/2)^2+(b+w)^2/4}+w/2-((N+1)w+Nb)}{(N+1)w+Nb}\right\}^2,$$

$$T_V\left\{\frac{(N-1)\sqrt{l^2+(b+w)^2}+\sqrt{l^2+(b+w)^2/4}+\sqrt{(l+b/2)^2+(b+w)^2/4}+w+b/2-(l+b+2w)}{l+b+2w}\right\}^2 \quad (1)$$

The bulk, Ω_B, and surface, Ω_S, porosities of the labyrinth structures are, respectively,

$$\Omega_B=\{[b(Nl+w)+(N-1)(2b+w)]/[((N+1)w+Nb)(l+b+2w)]\},\ \Omega_S=b/(l+b+2w),$$

$$\Omega_B=\{[(N+1)(l+b+w)b]/[((N+2)w+(N+1)b)(l+b+2w)]\},\ \Omega_S=b/((N+2)w+(N+1)b) \quad (2)$$

The complex density, ($\rho_L(\omega)$), complex compressibility, ($C_L(\omega)$), propagation constant, ($k(\omega)$), and normalised impedance ($Z_L(\omega)$) of an effective fluid medium equivalent to such a labyrinth structure are, respectively,

$$\rho_L(\omega)=(T/\Omega_B)\rho(\omega),\ \rho(\omega)=\rho_0/H(\lambda),\ H(\lambda)=1-\tanh(\lambda\sqrt{-i})/(\lambda\sqrt{-i}),\ C_L(\omega)=\Omega_B C(\omega),$$
$$C(\omega) = (\rho_0 c_0^2)^{-1}[\gamma-(\gamma-1)H(\lambda\sqrt{(N_{PR})})],\ \lambda=(b/2)\sqrt{\omega\rho_0/\mu},\ k_L(\omega)=\omega\sqrt{\rho_L(\omega)C_L(\omega)},$$
$$Z_L(\omega)=(\Omega_S\rho_0 c_0)^{-1}\sqrt{\rho_L(\omega)/C_L(\omega)}. \quad (3)$$

$\rho_0, c_0, N_{PR} \wedge \mu$ are density, sound speed, Prandtl number and dynamic viscosity in air, respectively.

3 Labyrinthine slit perforations in a rigid-porous matrix

With sufficient scale separation between matrix pore size and labyrinthine slit width, a perforated porous matrix may be regarded as a dual porosity system. The resulting complex density, ρ_{dpL}, and complex compressibility, C_{dpL}, are [3]

$$\rho_{dpL}(\omega)=\left[\frac{1}{\rho_L(\omega)}+\frac{(1-\Omega_B)}{\rho_m(\omega)}\right]^{-1},\ C_{dpL}(\omega)=C_L(\omega)+(1-\Omega_B)F_d(\omega)C_m(\omega), \quad (4)$$

where ρ_m and C_m represent the complex density and compressibility of the rigid-porous matrix and $F_d(\omega)$ is a pressure diffusion function.

Figures 2(a) and (b) show predictions of normal incidence absorption spectra for 5 cm thick hard-backed layers. If made in a non-porous rigid solid structure, labyrinthine slit perforations are predicted to yield small narrow absorption peaks (see the broken black lines in Figures 2(a) and (b)). But when labyrinths are made in a matrix with high flow resistivity and tortuosity, more useful low-frequency absorption peaks are predicted. Predictions in Figs. 2(a) and (b) are for a slanted slit matrix [4] perforated by an eight-fold horizontal slit labyrinth and a ten-fold vertical wall labyrinth respectively. The peaks at 260 Hz and 170 Hz in Figs. 2(a) and 2(b) correspond to the layer thickness being 1/26th and 1/40th of the incident wavelengths, respectively.

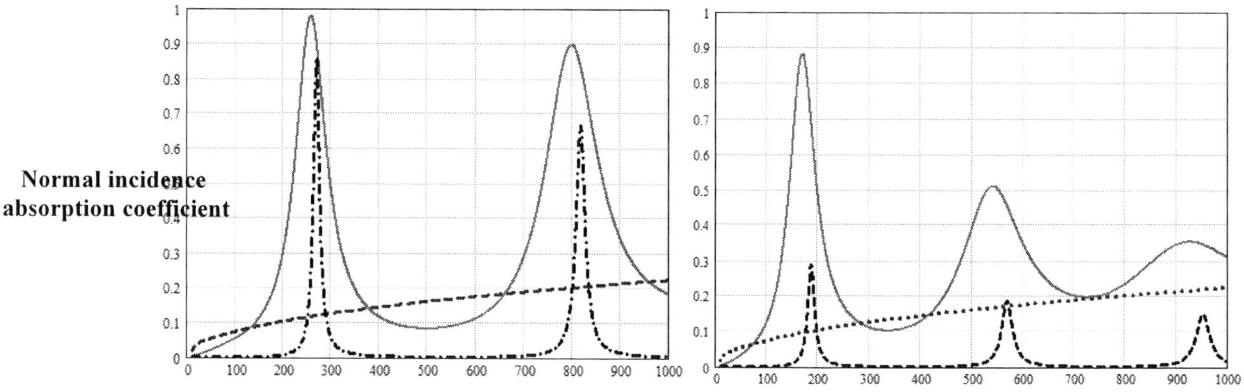

Figure 2: Normal incidence absorption coefficient spectra (red continuous lines) predicted for 0.05 m thick hard-backed layers consisting of (a) a horizontal wall labyrinth (w = 1 mm, b = 5.1 mm, l = 40 mm, N = 8) (b) a vertical wall labyrinth (w = 1 mm, b = 3 mm, l = 45 mm, N = 10) in an inclined slit medium (0.04 mm wide slits at 70° to the normal, porosity 0.9).
Black broken lines are predictions for corresponding labyrinthine perforations in a non-porous solid.
Blue dotted lines are predictions for each unperforated porous layer.

References

[1] W. Tang and C. Ren, Total transmission of airborne sound by impedance-matched ultra-thin metasurfaces, J. Phys. D: Appl. Phys. **50** 105102 (2017).

[2] A. Krushynska, *et al*, Labyrinthine Acoustic Metamaterials with Space-Coiling Channels for Low-Frequency Sound Control, Acta Acustica united with Acustica, **104** 200 – 210 (2018).

[3] X. Olny, C. Boutin, Acoustic wave propagation in double porosity media, J. Acoust. Soc. Am. **114** 73–89 (2003)

[4] K. Attenborough, Microstructures for lowering the quarter wavelength resonance frequency of a hard-backed rigid-porous layer, Applied Acoustics, **130** 188–194 (2018).

Morphology versus acoustic features of permeo-elastic materials

C. Boutin[1], R. Venegas[2]

[1] ENTPE Université de Lyon UMR CNRS 5513, Rue Maurice Audin, 69518 Vaulx-En-Velin, France
[2] University Austral of Chile, Institute of Acoustics, P.O. Box 567, 5090000 Valdivia, Chile

1 Introduction

This article focuses on the acoustic properties of porous media featuring a rigid solid skeleton with attached highly-flexible elastic films.This type of media can correspond, for example, to foams, with the thin membranes playing the role of films and the rigid structure being formed by their thicker struts. Such permeo-elastic media are characterized by the pore-scale fluid/film interaction effect which, compared to conventional porous media, enriches the flow physics with an elastic component. For this reason, the acoustic characteristics of permeo-elastic media can differ considerably from those of conventional porous media, see e.g. [1], [2].
However, such materials may have different morphologies depending on whether the pores or membranes are partially or fully connected. In this article it is shown that these morphological aspects play an essential role on the acoustic properties of these materials. We will successively examine the cases of permeo-elastic materials with pores that are: :
- Fully connected, i.e. no pores are closed by the films,
- Unconnected, i.e. all pores are closed by films,
- Partially connected, i.e. some pores are closed by the films.

To this end, by assuming acoustics wavelenghts much larger than the characteristic pore size, the two-scale asymptotic homogenization method have been used. The derived upscaled models enables to identify the different wave propagation regimes. In the following we will mainly present the assumptions and the results to which they lead. The theoretical developments will be reported in a forthcoming publication.

2 Fully connected pores

This case was investigated in [1], [2]. The main aspects are now recalled. Due to the separation of the scales and the fact that the pores are fully connected, the pressure at the dominant order is uniform in the pores, i.e. $P = P^0(x)$. The gas flow as well as the film movements are forced by the macroscopic pressure gradient. The local velocity field (of the gas and the films) results from the effects of the gas viscosity, the elasticity of the films (treated as bending plates and/or as tensioned membranes) and the inertia of the gas and the films. Thus, in a permeo-elastic medium, the constitutive law of flow which replaces the dynamic Darcy's law is of an elasto-visco-inertial nature. Moreover, the macroscopic mass balance is not altered by the thin films and is formally the same as in classical porous media (including thermal effects). For these materials three different flow regimes can be distinguished:

- At low frequencies, the films tend to behave in a quasi-rigid way and the flow is close to that of the porous medium (MP_0) for which the films would form part of the rigid skeleton. Nevertheless, the correction due to the films results in an elastic component which is added *in series* to the dynamic permeability of (MP_0),
- At high frequencies, the films tend to behave as if they were infinitely flexible and the flow is close to that of the porous medium (MP_∞) in the absence of films, but with an effective fluid/film mass. The correction provided by the films results in an elastic component which is added *in parallel* to the dynamic permeability of (MP_∞),
- At intermediate frequencies, the elastic effects become significant and combine with viscous and inertial effects. In particular, weakly damped internal resonance and anti-resonance regimes may develop. This situation is of interest for acoustics because in this frequency range the medium is quasi-lossless with a negative effective density that results in non-propagative waves.

3 Unconnected pores

In this case, although all pores are closed, *in accordance with the long wavelength hypothesis*, the pressure at the dominant order fluctuates on a macroscopic scale, i.e. $P = P^0(x)$. Thus, the developments are formally the same as in the previous case and the law of flow is also of an elasto-visco-inertial nature. Furthermore, the mass balance takes the same form as in classical porous media. However, as a consequence of the fact that the pores are closed, the average flow in the pores is entirely determined by the movement of the films. Consequently, the different flow regimes are as follows :

- At low frequencies, since films tend to behave in a rigid manner, the flow tends to fade and the medium to be impermeable. However, the macroscopic pressure gradient sets the films in motion and therefore the constitutive law of flow is dominated by the elasticity (and inertia) of the film. This *"elastic permeability"* is corrected by the visco (inertial) effect in the gas. The acoustic characteristics of the waves are therefore similar to those of weakly damped and weakly dispersive viscoelastic waves, and differ significantly, qualitatively and quantitatively, from those of low-frequency acoustic waves in conventional porous media.
- At intermediate frequencies, the elastic effects become strong and weakly damped; the internal resonance and anti-resonance of the gas cells coupled to the films appear, resulting in negative density and non-propagative waves.
- At high frequencies, since the films appear to be infinitely flexible, the flux is close to that of the porous medium (MP_∞) in the absence of films (again with a fluid/film effective mass). As in the case of connected pores, the presence of films introduces an elastic component which is added in parallel to the dynamic permeability of (MP_∞). This high frequency description applies only in the frequency range of scale separation. Above this range, internal dynamic phenomena (as local "breathing" modes) involving two or more cells may appear and leads to a more complex physics of acoustic waves.

4 Partially connected pores

In this morphology the pressure at the dominant order in the connected pores fluctuates at the macroscopic scale i.e. $P = P^0(x)$. In pores closed by films another pressure field P' can prevail, and the difference between the two pressures is balanced by the elasticity of the films. However, at the leading order, to respect the scale separation and the condition of local incompressibility the motion of the film must necessarily be of a lower order than that of the gas. Hence, at the dominant order the films are immobile. Thus, the gaz flow is identical to that which would develop in a rigid porous medium with the connected pores only. In this case we find the Darcy's classical visco-inertial law.
On the other hand, the contribution of the closed pores intervenes in the mass balance, and brings a correcting term to the effective compressibility. The said correction results from the elasto-inertial (and weakly dissipative by viscosity) behavior of the resonator constituted by the film and volume of the trapped gas.
Thus at low and high frequencies we find an usual behavior. Conversely in the local resonance frequency range an unconventional dispersion of acoustic wave can be observed as the effective compressibility is significantly modified and may even become negative.

5 Conclusion

In conclusion, while it is well known that the morphology plays a determining role in porous media, it seems its influence is even more important for permeo elastic media, for which they can lead to a drastic qualitative and quantitative change of acoustic behavior.

References

[1] R. Venegas, C. Boutin, *Acoustics of permeo-elastic materials*, J. Fluid Mech. (2017)

[2] C. Boutin, R. Venegas, *Pore-scale bending and membrane effects in permeo-elastic media*, Mech. of Mat. (2020)

Energetic homogenization of one-dimensional porous material with variable cross-sectional area

M. B. Muhlestein

US Army Engineer Research and Development Center, 72 Lyme Rd.,
Hanover, NH 03755

A recently developed energetic homogenization method [Muhlestein, J. Acoust. Soc. Am. **5***, 3584-3593 (2020)] is used to estimate the apparent material properties of a simple porous material with varying cross-sectional areas. Expressions for the apparent mass density, compressibility, and kinematic viscosity are given. As a demonstration, a simple duct with identical expansion chambers is briefly considered.*

1 Introduction

This paper is concerned with the homogenization of a porous material that may be modeled as a statistically identical network of one-dimensional ducts with varying cross-sectional areas. For this paper the ducts are considered to be acoustically wide, such that boundary layers may be assumed to be thin compared to the duct width. Using a recently developed energetic homogenization technique, described below in Sec. 2, the impact of the varying cross-sectional area on the apparent mass density, bulk modulus, and kinematic viscosity may be calculated. A demonstration of the technique using a simple expansion chamber is given in Sec. 3.

2 Energetic Homogenization

One approach to homogenization is to average the weak form of the dynamic equations, known as Hamilton's principle, by assuming that the acoustic pressure and volume velocity are constants throughout a representative volume element [1]. Hamilton's principle requires knowledge of the total kinetic and potential energies, as well as any constraints and external forces (such as from friction). For a one-dimensional duct with variable cross-section we may write the kinetic and potential energies, respectively, as

$$E_k = \int_0^L \frac{1}{2S(x)}\rho_0 q^2 \, dx, \qquad E_p = \int_0^L \frac{S(x)}{2}\beta_0 p^2 \, dx, \tag{1}$$

where the domain $x \in (0, L)$ is the extent of the duct of interest, $S(x)$ is the cross-sectional area, ρ_0 is the fluid mass density, β_0 is the fluid compressibility (inverse bulk modulus), q is the volume velocity, and p is the acoustic pressure. The acoustical fields are assumed to be functions of position x and time t. The continuity equation may be imposed via the constraint equation $\mathcal{C} = 0$ and boundary layer losses may be incorporated as an external force density F [2], where

$$\mathcal{C} \equiv S(x)\beta_0 \frac{\partial p}{\partial t} + \frac{\partial q}{\partial x} = 0, \qquad F = -\rho_0\sqrt{\nu_0}\frac{C(x)}{S^2(x)}\frac{\partial^{1/2} q}{\partial t^{1/2}}, \tag{2}$$

where ν_0 is the kinematic viscosity that depends only on the propagation fluid, $C(x)$ is the duct perimeter at position x, and $\partial^{1/2}/\partial t^{1/2}$ is the half time derivative operator (see, e.g., Ref. [3]). Combining these expressions, we may write Hamilton's principle as

$$\delta \int_{t_0}^{t_1}\int_0^L \left[\frac{\rho_0}{2S(x)}q^2 - \frac{S(x)\beta_0}{2}p^2 + \mu\mathcal{C}\right] dx\, dt = \int_{t_0}^{t_1}\int_0^L \rho_0\sqrt{\nu_0}\frac{C(x)}{S(x)}\frac{\partial^{1/2} q}{\partial t^{1/2}}\delta u \, dx\, dt, \tag{3}$$

where t_0 and t_1 are arbitrary initial and final times, δ is the variational operator, μ is a Lagrange multiplier function for the constraint equation, and u is the volume displacement (such that $q = \partial u/\partial t$). Note that in this form of Hamilton's principle the only acoustical field quantities that appear are those that are continuous across interfaces (i.e., the acoustic pressure and various time derivatives or integrals of the volume displacement). Then, following Muhlestein [1], we may conclude the macroscopic acoustical quantities behave as though the duct had constant (and chosen arbitrarily) cross-sectional

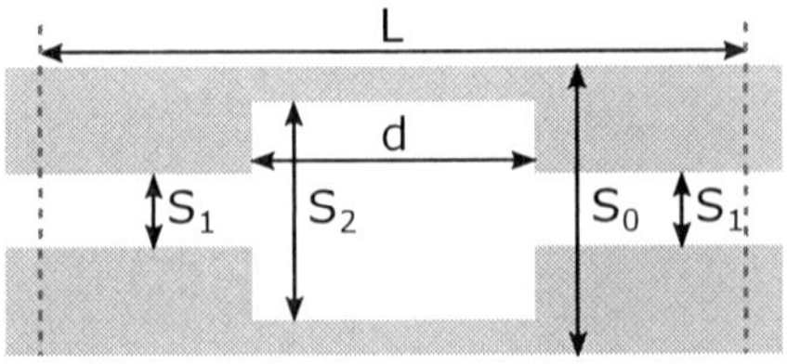

Figure 1: Schematic of an example duct with a variable cross-sectional area, as well as the reference cross-sectional area S_0.

area S_0 and perimeter C_0, and the propagation fluid had effective mass density, compressibility, and boundary drag coefficient respectively given by

$$\rho_{\text{eff}} = \rho_0 S_0 \left\langle \frac{1}{S} \right\rangle, \qquad \beta_{\text{eff}} = \beta_0 \frac{\langle S \rangle}{S_0}, \qquad \rho_{\text{eff}} \sqrt{\nu_{\text{eff}}} = \rho_0 \sqrt{\nu_0} \frac{S_0}{C_0} \left\langle \frac{C}{S} \right\rangle. \tag{4}$$

From these quantities one may derive an effective wavenumber, acoustic impedance, and absorption values.

While this presentation is limited to considering boundary layer losses in linear media, bulk thermoviscous losses and nonlinear phenomena may be readily incorporated (see Ref. [1]). Additional geometric features (e.g., dead-end pores) or physical phenomena may also be accounted for using this formalism.

3 Demonstration

As a demonstration of the homogenization method, consider the case of an air-filled duct with simple cylindrical expansion chambers (see Fig. 1). Let the reference cross-sectional area be S_0, the cross-sectional area of the narrow tubes be S_1, and the cross-sectional area of the expansion chambers be S_2, the length of the element is L and the length of the expansion chamber is d. Using Eqs. (4) we may then write

$$\frac{\rho_{\text{eff}}}{\rho_0} = \frac{S_0}{L} \frac{S_1 d + S_2(L-d)}{S_1 S_2}, \qquad \frac{\beta_{\text{eff}}}{\beta_0} = \frac{S_1(L-d) + S_2 d}{L S_0}, \qquad \sqrt{\frac{\nu_{\text{eff}}}{\nu_0}} = \frac{C_2 S_1 d + C_1 S_2 (L-d)}{C_0 S_1 d + C_0 S_2 (L-d)}, \tag{5}$$

where C_i is the circumference associated with S_i. Since the porosity may be written as $\phi = [S_1(L-d) + S_2 d]/L$, the structure factor may be written as

$$k_s = \phi \frac{\rho_{\text{eff}}}{\rho_0} = \frac{[S_1(L-d) + S_2 d]\,[S_2(L-d) + S_1 d]}{L^2 S_1 S_2}, \tag{6}$$

in agreement with Neithalath, et al. [4].

It is possible that more extreme effective properties could be obtained by incorporating more complicated geometries, such as dead-end resonators or membranes.

References

[1] M. B. Muhlestein. "Effective Acoustic Metamaterial Homogenization Based on Hamilton's Principle with a Multiple Scales Approximation". *J. Acoust. Soc. Am.* 147.5 (2020), 3584–3593.

[2] D. T. Blackstock. *Fundamentals of Physical Acoustics.* Hoboken, NJ: John Wiley & Sons, Inc., 2000.

[3] N. Sugimoto. "Burgers Equation with a Fractional Derivative; Hereditary Effects on Nonlinear Acoustic Waves". *J. Fluid Mech.* 225 (1991), 631–653.

[4] N. Neithalath, A. Marolf, J. Weiss, and J. Olek. "Modeling the Influence of Pore Structure on the Acoustic Absorption of Enhanced Porosity Concrete". *J. Adv. Concrete Tech.* 3.1 (2005), 29–40.

A knowledge of the surface impedance of a resonant surface backing a fibrous layer can be insufficient when calculating the absorption of the combined system

Ronald Gerdes, Thomas Herdtle

3M Company, 3M Center, St. Paul, MN, USA

1 Introduction

Complex structures, such the acoustically absorbing honeycomb (AAH) structure [1] shown in Fig. 1, can be modeled in full 3D with the thermo-viscous acoustics module in COMSOL Multiphysics [2] to obtain the absorption as well as the complex surface impedance. The latter can then be used in a simpler 1D model of the impedance tube, where the surface impedance is used as the tube's boundary condition instead of explicitly modeling the actual structure. The same could be done with a fibrous layer, first modeling its full thickness, and then replacing that fiber domain with its surface impedance; the equivalence is shown in Fig. 2.

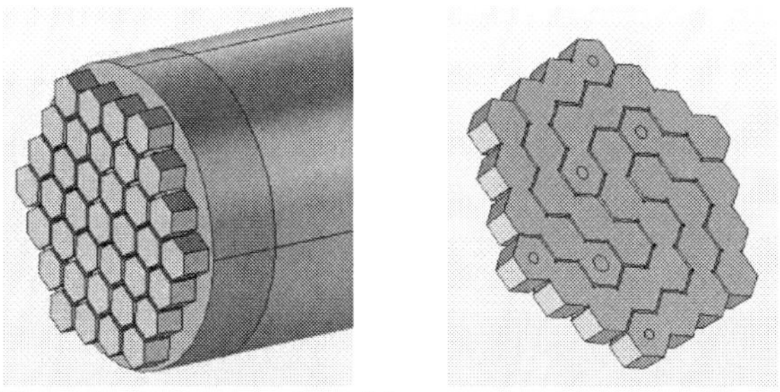

*Figure 1: **Left** – Virtual 100 mm impedance tube with complex-connected honeycomb structure at the end. The blueish region is initially an air domain and is later replaced with a fibrous layer. **Right** – The interconnected hexagons form channels of various lengths with various openings.*

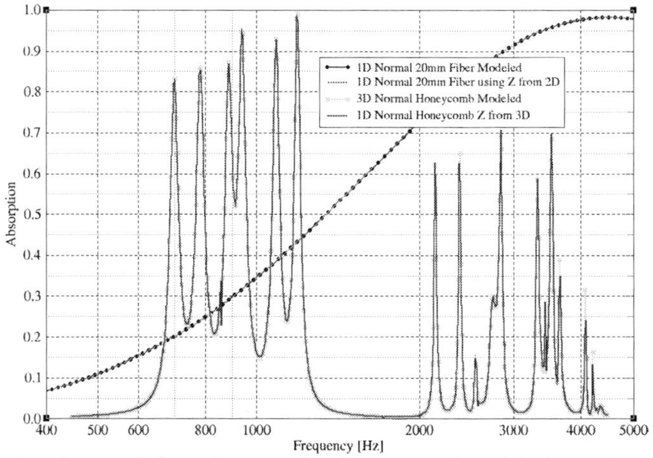

Figure 2: Demonstration that modeling the complex honeycomb in 3D (green) can be replaced with the surface normal impedance in a 1D model (purple). The same can be done for a fibrous layer.

2 Modeling Results

However, care must be taken when using the surface impedance in more complex, multilayer models. For example, it has been found that when trying to model a fibrous layer in 1D, overlaying a surface whose normal impedance was determined from the 3D thermo-viscous AAH model, the results do not match those of the full 3D model where the fiber layer is included along with the explicitly-modeled 3D structures. In both cases the fibrous layer was modeled as a complex fluid using the Delany-Bazley formulation, with its 20 mm depth.

As can be seen in Fig. 3, the full 3D model (dark blue) shows the individual peaks and valleys are completely averaged out and the absorption curve is very smooth. On the other hand, when the surface impedance of the

AAH is used on the back side of the fibrous layer (light blue), the peaks are almost identical to the untreated sample, but the troughs are less deep. These dramatically different results prompted us to perform some experimental verification.

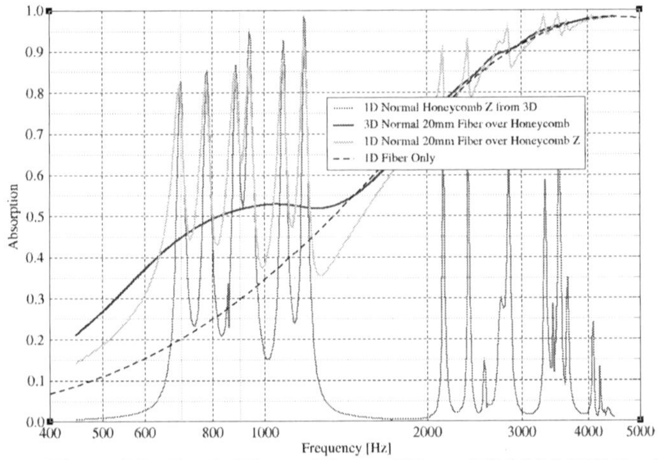

Figure 3: ***Red** – AAH from 3D model.* ***Dark Blue** – 20 mm Fiber. 20,000 MKS-Rayl/m, over 3D AAH model.* ***Light Blue** – 20 mm Fiber over the AAH surface impedance.* ***Black Dashed** – 20 mm Fiber only.*

3 Experimental Results

A 3D printed version of AAH, with 4 resonances, was tested in a 100 mm impedance tube, with and without an overlaying 20 mm fibrous layer with a flow resistivity of approximately 15,000 MKS-Rayl/m. As seen in Fig. 4, the behavior of the AAH with a top fiber layer (green curve) behaves very similarly to the full 3D AAH model with fiber.

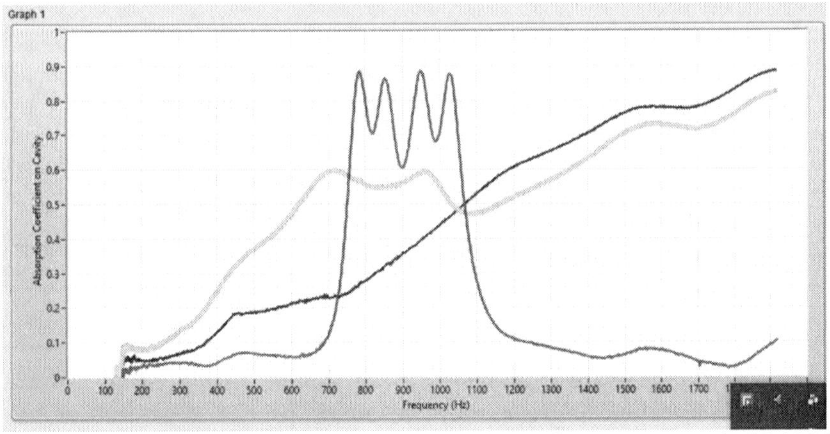

Figure 4: ***Red** – AAH test article only.* ***Blue** – 20 mm Fiber only, approximately 15,000 MKS-Rayl/m.* ***Green** – 20 mm Fiber over the AAH test article.*

4 Conclusion

One cannot take the normal surface impedance of a complex structure and add layers of other materials, e.g., represented by complex fluids, to obtain the absorption of the composite structure. Instead, a full model of the materials, including the structure, is required to obtain the true absorption. This is likely due to the near-field hole-to-hole interactions, which are significantly affected by the presence of the fiber layer, are completely eliminated when only using the apparent far-field surface impedance of the structure.

5 References

[1] J. Jonza, T. Yoo, T. Herdtle, J. Kalish, R. Gerdes, G. Eichhorn, *Acoustically Absorbing Lightweight Thermoplastic Honeycomb Panels*, INTER-NOISE and NOISE-CON Congress and Conference Proceedings 254 (2), 210-219.

[2] COMSOL Multiphysics v. 5.5. www.comsol.com. COMSOL AB, Stockholm, Sweden.

A vibration transmissibility dynamic poroelastic model for open cell polyurethane foam involving viscoelasticity and pneumatic damping effects

Qicheng Zhang[1], Fabrizio Scarpa[1*], David Barton[2], Yunpeng Zhu[3], Ziqiang Lang[3]

[1]Bristol Composites Institute (ACCIS), University of Bristol, BS8 1TR Bristol, UK
[2]Department of Engineering Mathematics, University of Bristol, BS8 1UB Bristol, UK
[3]Department of Automatic Control and Systems Engineering, University of Sheffield, Sheffield, UK

This paper presents a dynamic poroelastic model for polyurethane (PU) foam considering the viscoelasticity of the PU material and the pneumatic damping of the porous structure. Stress relaxation, quasi-static cyclic loading and vibration transmissibility tests have been performed on PU foam samples to identify viscoelastic and permeability parameters.

1 Introduction

PU foam is widely used for vibration or noise absorption. The high damping of the foam is mainly caused by the viscoelasticity of the PU material and the pneumatic damping effect associated to the porosity and dynamic permeability. Bianchi and Scarpa [1] have previously developed dynamic models for PU conventional and auxetic foam considering only the pneumatic damping phenomenon. White et al [2] have also developed dynamic models of PU foam focusing on the viscoelasticity and ignoring the pneumatic damping effect. In this work we have modified the pneumatic damping model to also consider more realistic air flow perpendicular to the loading direction and including the intrinsic viscoelasticity of the polyurethane. Hereditary integral models based on Prony series [3] are widely used to describe viscoelasticity. We have here developed a method to consider the stress relaxation test and quasi-static cyclic loading test data simultaneously to overcome known identification problems of viscoelastics [2] and determine the Prony series parameters for the viscoelastic material based on a nonlinear regression method.

2 Methodology and results

The Reynolds number within the internal airflow in the open cell foam is estimated to be around 0.1 in our testing conditions, thus the linear Darcy's law can be applied. The air is assumed to be incompressible and the air flow along the direction of vibration is ignored because the layout of the vibration transmissibility rig [1]. The relation between the pneumatic damping force and deformation rate of the foam can be described in equation (1), where μ is the dynamic viscosity of the air, φ is porosity of the foam, K is permeability, W=30mm is width of the foam sample and H=15mm is height of the sample.

$$F_{a-z}(\dot{z}) = \frac{W^4 \mu}{32\varphi K H}\dot{z} \tag{1}$$

Stress-time curves of the PU in stress relaxation and cyclic loading tests are shown in equation (2) and (3), obtained from hereditary integral model based on Prony series. Fourier series with order M=20 are used to describe the triangle wave during the cyclic loading. E_0 is modulus (linear within small strain ranges up to <2%). a_i and α_i are parameters of Prony series with order N=3.

$$\sigma_1(t) = E_0\varepsilon_0\left[1-\sum_{i=1}^{N}\frac{a_i}{\alpha_i}\left(1-e^{-\alpha_i t}\right)\right] \tag{2}$$

$$\begin{aligned}\sigma_2(t) = {} & \frac{8E_0\varepsilon_{\max}}{\pi^2}\sum_{k=0}^{M-1}\left[\frac{(-1)^{k+1}}{(2k+1)^2}\sin\left((2k+1)\omega_0 t\right)\right] - \\ & \frac{8E_0\varepsilon_{\max}}{\pi^2}\sum_{i=1}^{N}\sum_{k=0}^{M-1}\left\{\frac{(-1)^{k+1}\omega_0 a_i}{(2k+1)\left(\alpha_i^2+(2k+1)^2\omega_0^2\right)}\left[-\cos\left((2k+1)\omega_0 t\right)+\frac{\alpha_i}{(2k+1)\omega_0}\sin\left((2k+1)\omega_0 t\right)+e^{-\alpha_i t}\right]\right\}\end{aligned} \tag{3}$$

$$\chi^2(\mathbf{p}) = \sum_{i=1}^{N_1}\left[\sigma_{1i}-\sigma_1(t_{1i},\mathbf{p})\right]^2 + \beta\sum_{j=1}^{N_2}\left[\sigma_{2j}-\sigma_2(t_{2j},\mathbf{p})\right]^2 \tag{4}$$

The error function is shown in equation (4), where σ_{1i}, t_{1i}, σ_{2j} and t_{2j} are the experimental data from the stress relaxation and cyclic loading tests, β=2.5 is the weight factor and **p** are viscoelastic parameters. The

Marquardt-Levenberg Method is used to minimize the error and obtain the viscoelastic parameters (Table 1). The comparison between experimental and simulated curves related to stress relaxation and cyclic loading tests (strain rate $\dot{\varepsilon}$=0.00056/s) are in Figure 1 (a) and (b), showing good agreement.

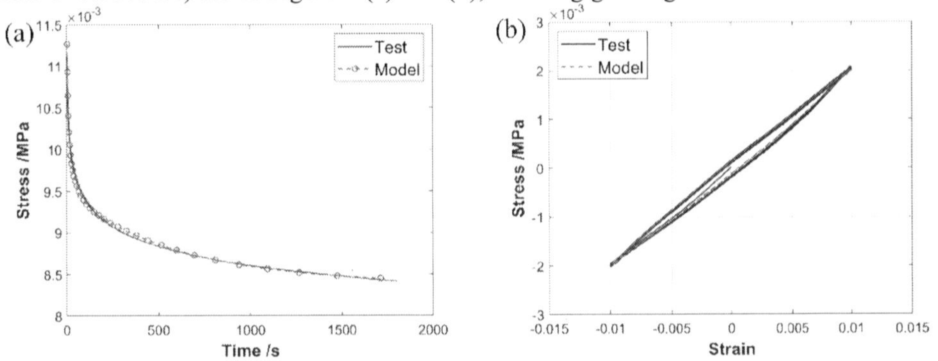

Figure 1 Stress relaxation test result (a) and quasi-static cyclic loading test result (b)

Symbol	E_0 /MPa	a_1 /s^{-1}	a_2 /s^{-1}	a_3 /s^{-1}	α_1 /s^{-1}	α_2 /s^{-1}	α_3 /s^{-1}
Value	0.218958	0.000152	0.016686	0.001277	0.001634	0.128591	0.022067

Table 1 Viscoelastic parameters of the PU foam from inverse.

The dynamic model of the transmissibility test rig for PU foam specimens is in equation (5), where M=116g is top mass, m_f=0.38g is foam mass, $u_d(t)$ is foam deformation and $u_b(t)$ is the base displacement. The transfer function (TF) of the system can be obtained from equation (5) using Fourier transform. Experimental TF curves using white noise and sine sweep are in Figure 2 (a) and show good agreement. The permeability of the foam is then obtained by inverse identification from the experimental TF curves (Figure 2 (b)). K increases almost linearly with the frequency: the pneumatic damping force is almost constant with frequency.

$$\left(M+\frac{m_f}{3}\right)\ddot{u}_d(t)+\frac{W^4\mu}{32\varphi KH}\dot{u}_d(t)+\frac{W^2}{H}\left[E_0u_d(t)-E_0\int_{-\infty}^{t}\sum_{i=1}^{N}a_ie^{-\alpha_i(t-\tau)}u_d(\tau)d\tau\right]=-\left(M+\frac{m_f}{2}\right)\ddot{u}_b(t) \qquad (5)$$

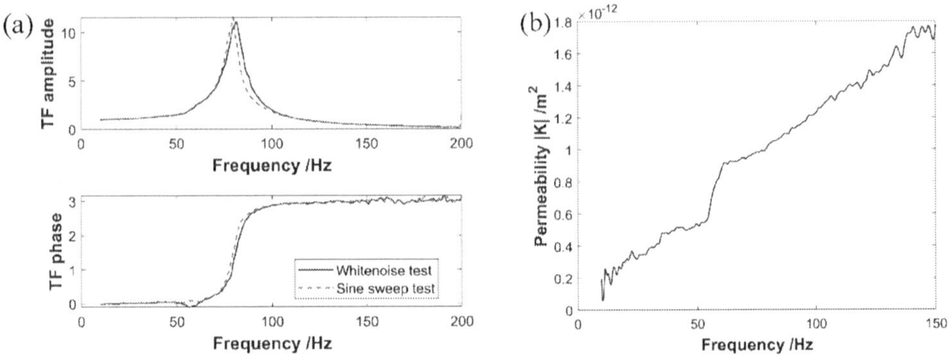

Figure 2 Transfer function of foam specimen in vibration test (a) and permeability identified (b)

3 Conclusions

A new dynamic poroelastic model for the vibration transmissibility of PU foams includes the pneumatic damping effect of the porosity, permeability and viscoelasticity of the PU material. Viscoelastic parameters are obtained by stress relaxation test and quasi-static cyclic loading test data simultaneously. The permeability of the foam at low frequencies is obtained by inverse identification from experiments.

This project is supported by the UK Engineering and Physical Sciences Research Council (EPSRC) EP/R032793/1 SYSDYMATS.

References

[1] M. Bianchi, and F. Scarpa. *Vibration transmissibility and damping behaviour for auxetic and conventional foams under linear and nonlinear regimes.* Smart Mater. Struct. 22.8 (2013): 084010.

[2] W. S. White, et al. *Experimental techniques and identification of nonlinear and viscoelastic properties of flexible polyurethane foam.* Nonlinear Dyn 22.3 (2000): 281-313.

[3] C. Tzikang. *Determining a Prony series for a viscoelastic material from time varying strain data.* (2000) NASA / TM-2000-210123.

Extension of frequency range of the eight-microphone method in normal-incidence sound absorption coefficient measurement

A. Sanada[1], H. Nakagawa[2]

[1] Industrial Technology Center of Okayama prefecture
5301 Haga, Kita-ku, Okayama-shi, Okayama 701-1296, Japan
[2] Nihon Onkyo Engineering Co., Ltd.
1-21-10 Midori, Sumida-ku, Tokyo, Japan

1 Introduction

This study relates to a method for further increasing the upper limit frequency for measuring the normal-incidence sound absorption coefficient using eight microphones. With the conventional measurement method (so-called 2-microphone transfer function method [1]), the normal-incidence sound absorption coefficient can be measured when only the (0,0) mode can propagate in the impedance tube. By placing four microphones at 90 degree intervals on the same cross section and placing the tips of the microphones at the node of the radial (0,1) mode, it is possible to eliminate the influence of the circumferential (1,0) (2,0) (3,0) modes and the radial (0,1) mode. This makes it possible to measure the normal incident sound absorption coefficient up to a frequency band that is about three times higher than the conventional measurable upper limit frequency.

2 Method

The sound field inside the cylindrical tube with radius *R*, the acoustic modes generated in the cross-sectional direction inside the impedance tube are as shown in Figure 1. *m* and *n* represent the mode orders in the circumferential and radial directions of the impedance tube cross section respectively. λ_{mn} is a value of *kR* that satisfies the boundary condition equation 1 on the wall surface of the impedance tube.

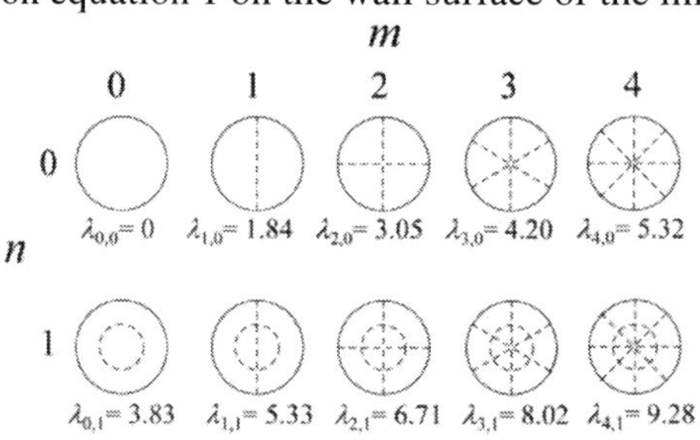

Figure 1 : Acoustic modes in the tube

$$\left.\frac{dJ_m(k_{r(m,n)}r)}{dr}\right|_{r=R} = 0. \qquad [1]$$

In the two-microphone transfer-function method, measurement is conducted under conditions where only the (0, 0) mode can propagate in the tube. Thus, the measurable frequency in the conventional method is below the cut-on frequency of the (1,0). As the frequency increases beyond that frequency, each high-order mode starts to propagate at its cut-on frequency. When these higher-order modes exist in the tube, the normal-incidence absorption coefficient cannot be measured by the two-microphone transfer-function method because of the complexity of the sound field.

The authors utilize the circumferential mode characteristics of the impedance tube to place four microphones on the same cross section at 90 degree intervals. By summing the sound pressure signals of the four microphones, the (1,0) (2,0) (3,0) modes are cancelled. Also, by placing the microphones at the nodal line of the radial (0,1) mode, the effects of this radial (0,1) mode is eliminated. This makes it possible to measure only the sound waves propagating in the axial direction of the impedance tube even within the frequency range in which the (1,0) (2,0) (3,0) and (0,1) modes propagate and enable to estimate normal-incidence absorption coefficient.

3 Measurement

Figure 2 shows an image of the impedance tube and microphone arrangement. The inner diameter of the impedance tube is 100 mm. Four microphones are placed at 90 degree intervals on two cross sections, and the tip of each microphone is adjusted to the nodal line of the (0,1) mode. Figure 3 shows a photograph of the actual impedance tube.

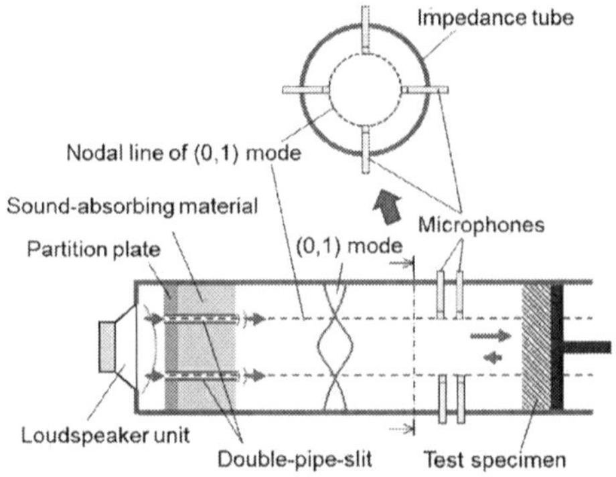

Figure 2

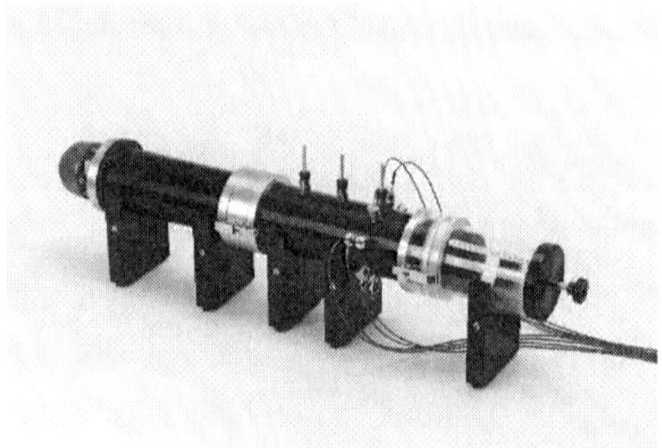

Figure 3

4 Results

Measured samples are Melamine foam with a thickness of 25 mm and urethane foam with a thickness 20mm. The measurement results are shown in Fig. 4. The red solid lines are the sound absorption coefficient measured by this development method (8 microphone method). The green solid lines are the sound absorption coefficient measured by the conventional method (2 microphone method) with an acoustic tube with a diameter of 40 mm.

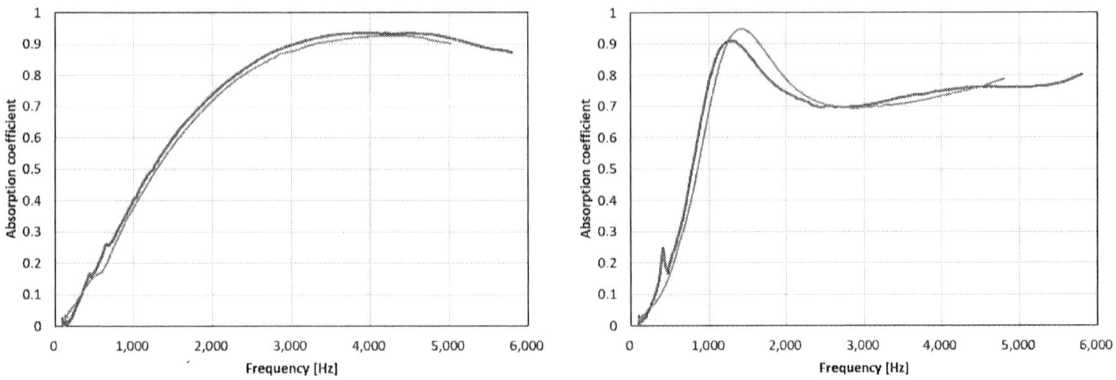

Figure 4 : Measurement results. Red line is φ100 tube 8-microphone method, green line is φ40 tube conventional 2-microphone method. Left – Melamine foam 25mm, right : urethane foam 20mm.

References

[1] ISO 10534-2, *Acoustics–Determination of sound absorption coefficient and impedance in impedance tubes –Part 2: Transfer-function method* (1998)

[2] A. Sanada, *Extension of the frequency range of normal incidence sound absorption coefficient measurement using four or eight microphones*, Acoustical Science and Technology, 38(5), 261-263 (2017).

[3] A. Sanada1, K. Iwata1 and H. Nakagawa, *Extension of the frequency range of normal-incidence sound absorption*
coefficient measurement in impedance tube using four or eight microphones, Acoustical Science and Technology, 39(5), 335-342 (2018).

Determination of angle-dependent absorption coefficients of porous materials based on the modal decomposition method

Z. Zhang[1,3], H. Denayer[1,3], C. Claeys[1,3], W. Desmet[1,3], E. Deckers[2,3]

[1] KU Leuven, Dept. of Mechanical Engineering,
Celestijnenlaan 300B, 3001 Leuven, Belgium
[2] KU Leuven, Campus Diepenbeek, Dept. of Mechanical Engineering,
Wetenschapspark 28, 3590 Diepenbeek, Belgium
[3] DMMS lab, Flanders Make, Belgium

This paper presents a way to measure the angle-dependent absorption coefficients of porous materials in a rectangular waveguide. An optimization of microphone locations is made to make the method robust against the measurement noise. It is further validated experimentally.

1 Introduction

Impedance tubes are commonly used to determine the absorption coefficients of porous materials at normal incidence. The angle-dependent absorption coefficients can not be measured easily therein since only plane waves can propagate in the working frequency range of the impedance tubes [1]. Existing setups to measure the angle-dependent absorption coefficients are based on methods such as in-situ method [2] and spatial Fourier-transform method [3], both of which have drawbacks. The in-situ method suffers from requirement of large sample size. The spatial Fourier-transform method suffers from unrealistic long waveguide. In this paper, we propose to use the modal decomposition method to obtain angle dependent properties in a setup of a moderate size.

2 Methodology

The acoustic field inside a rigid rectangular waveguide can be regarded as a weighted sum of all cut-on modes [4]. As a mode can be regarded as acoustic plane waves of an oblique angle, the absorption coefficients of the corresponding oblique angles can be calculated from that of the modes. The absorption coefficient of each mode can be calculated by the scattering matrix, which is calculated by measuring the acoustic pressure at a sufficient number of points in the field for a sufficient number of independent sound source locations. If N is the total number of cut-on modes at a frequency, theoretically, at least $2(N+1)$ points need to be measured in the field and at least $2(N+1)$ independent measurements need to be performed.
An optimization is performed on the microphone locations to improve the matrix inversion stability when calculating the scattering matrix [5]. Figure 1 shows the in-house designed waveguide with microphone locations optimized between 1000 Hz and 3000 Hz.

Figure 1: The waveguide designed and optimized between 1000 Hz and 3000 Hz

3 Results and discussion

A foam sample with known parameters is measured in the designed waveguide to validate the setup. In Figure 2 (left) a comparison is shown between the measurement from the designed waveguide and the simulation of the angle-dependent absorption coefficients at 2000 Hz. In Figure 2 (right) a comparison is shown between the measurement from the designed waveguide and that from an impedance tube of reflection and transmission coefficients at normal incidence.

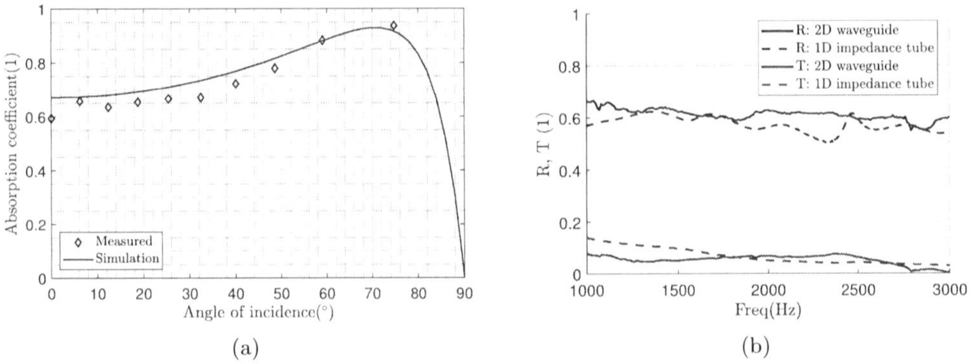

Figure 2: (a) The angle-dependent absorption coefficients at 2000 Hz from simulation and from designed waveguide measurement (b) The reflection and transmission coefficients of a foam sample under normal incidence measured in the designed waveguide and in an impedance tube

4 Conclusions

In this work, it is shown that the modal decomposition method can be used to determine angle-dependent absorption coefficients of a porous sample. The method has been experimentally validated.

Acknowledgments

The research of Z. Zhang is funded by a grant within the European Project PBNv2 Marie Curie Initial Training Network (GA 721615). The Internal Fund KU Leuven is gratefully acknowledged for its support.

References

[1] ISO. 10534-2: 2001. *Acoustics-Determination of sound absorption coefficient and impedance in impedance tubes-Part*, 2:10534–2, 2001.

[2] C Nocke. In-situ acoustic impedance measurement using a free-field transfer function method. *Applied Acoustics*, 59(3):253–264, 2000.

[3] Masayuki Tamura, Jean F Allard, and Denis Lafarge. Spatial fourier-transform method for measuring reflection coefficients at oblique incidence. ii. experimental results. *The Journal of the Acoustical Society of America*, 97(4):2255–2262, 1995.

[4] Mats Åbom. Modal decomposition in ducts based on transfer function measurements between microphone pairs. *Journal of Sound and Vibration*, 135(1):95–114, 1989.

[5] Ze Zhang, Hervé Denayer, Claus Claeys, Wim Desmet, and Elke Deckers. Angle-dependent reflection, transmission and absorption coefficients measurement using a 2d waveguide. *Applied Acoustics*, 177:107946, 2021.

The Test Principles of Biot Properties of Poro-Elastic Material and Its Engineering Validation

J. Pang[1], X. Xiong[1], H. Zhang[1]

[1] Room 322, Building 10, 658 Jinzhong Rd.,
PROSYNX Technolgy Inc., Shanghai, CHINA

As we well known, porous materials, because of its best absorption performance, have been widely used in the field of noise reduction. Its theoretical research is becoming more and more perfect, the most classic is Biot theory. The full poro-elastic Biot model for materials involves in the fluid and elastic properties. How to obtain these characteristic parameters correctly is the key for acoustic research. In this presentation we will introduce the basic test principle from the definition of Biot parameters and the applicable standards or methods. Finally, the correctness and practicability are verified by some engineering cases.

1 Introduction

In this presentation, I will explain how to obtain the parameters of bio model by experimental method from the following aspects:

- What about conventional porous materials in acoustics?
- Conventional theoretical models of porous materials.
- The definition of Biot parameters and related test standards and methods.
- Introduction of test equipment for Biot parameters.
- Engineering validation.

2 General information

2.1 The basic characteristics of porous materials

- Two phases: solid and fluid
- Elastic coupling
- Visco-inertial coupling
- Energy dissipation
  - viscous effect
  - thermal effect
  - structural damping

2.2 Macroscopic model

- 5-parameter Johnson-Champoux-Allard (JCA) model
- 6-parameter Johnson-Champoux-Allard-Lafarge (JCAL) Model
- 9-parameter Biot poro-elastic model

2.3 Biot parameters and related test standards and methods.

The popular parameters include: Open porosity, Static airflow resistivity, Tortuosity, Viscous characteristic length, Thermal characteristic length, Young's modulus, Poisson's ratio, Damping loss factor.

Parameters	*Definition*	*Standard*	*Method*
Porosity (ϕ)	$\phi = \frac{V_{fluid}}{V_{total}}$	N/A	Pressure/mass method
Bulk Density(ρ)	$\rho = \frac{m_s}{V_t}$	N/A	Pressure/mass method
Airflow Resistivity (σ)	$\sigma = \frac{\Delta P}{Q}\frac{A}{h}$	ISO 9053	Direct method

Tortuosity(α_∞)	$\alpha_\infty = \left(\frac{L'}{L}\right)^2$	N/A	Ultrasound method
Viscous characteristic length(Λ)	$\Lambda = 2\int_V v^2 dV \Big/ \int_S v^2 dS$	N/A	Ultrasound method
Thermal characteristic length(Λ')	$\Lambda' = 2\int_V dV \Big/ \int_S dS$	N/A	Ultrasound method
Young's modulus(E)		ISO 18437-5	Quasi-static compression test
Poisson's ratio(ν)		ISO 18437-5	Quasi-static compression test
Damping loss factor(η)		ISO 18437-5	Quasi-static compression test

Table 1 : Biot parameters and their test standards

2.4 Engineering validation

The comparison is given in figure 1. Very good agreement is observed.

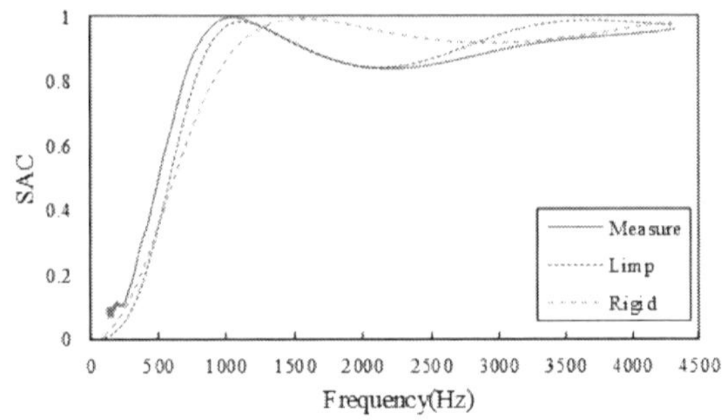

Figure 1 : The comparison between test and theoretical computation

References

[1] Allard J.F. & Atalla N., Propagation of sound in Porous media: Modelling Sound Absorbing Materials. 2e. John Wiley & Sons Ltd., 2009.

[2] INTERNATIONAL STANDARD, ISO 18437-5, Mechanical vibration and shock -- Characterization of the dynamic mechanical properties of visco-elastic materials -- Part 5: Poisson ratio based on comparison between measurements and finite element analysis, First edition, 2011.

[3] INTERNATIONAL STANDARD, ISO 9053-1, Acoustics – Determination of airflow resistance – Part 1: Static airflow method.

[4] Y. Salisssou and R. Panneton, "Pressure/mass method to measure open porosity of porous solids," J. Applied Physics 101, 124913, 1-7 (2007).

[5] J.-F. Allard, B. Castagnède, M. Henry, W. Lauriks, "Evaluation of tortuosity in acoustic porous materials saturated by air," Rev Sci. Instrum. 65, 754–755 (1994).

[6] Z.E.A. Fellah, S. Berger, W. Lauriks, C. Depollier, C. Aristégui, J.-Y. Chapelon, "Measuring the porosity and the tortuosity of porous materials via reflected waves at oblique incidence," J. Acoust. Soc. Am. 113, 2424-2433 (2003).

[7] N. Kino, "Ultrasonic measurements of the two characteristic lengths in fibrous materials," Applied Acoustics 68, 1427-1438, 2007.

A New Device for Fluid Equivalent Parameters Assessment

J.C. Le Roux[1], J.P. Dalmont[2], N. Poulain[1]

[1]Centre de Transfert de Technologies du Mans, Le Mans, France

[2]Laboratoire d'Acoustique de l'Université du Mans, UMR CNRS 6613, Le Mans Université, Le Mans, France

1 Introduction

Several experimental methods have been developed for the estimation of poro-elastic fluid equivalent parameters: low or high frequencies acoustical methods, non acoustical methods. Low frequencies acoustical methods [1, 2, 3,] are based on the measurements of the equivalent dynamic density and bulk modulus of the porous medium. They can be divided into analytical and inverse methods. The advantage of these methods is that they do not require specific or expensive equipment. However, most of these methods require prior knowledge or estimation of the static flow resistivity and open porosity.

In this paper, we present a new approach based on the use of an impedance sensor [4] with significant low frequency extension than classical duct methods. Dynamic density and bulk modulus are therefore measured over a wide frequency range, and the static air flow resistivity and open porosity can be directly obtained from the low frequency asymptotes (of the equivalent dynamic density and bulk modulus), an approach which is not possible with classical duct methods. The remaining fluid equivalent parameters can be analytically obtained from the same measurements using the well-known analytical methods [1,2].

2 Description of the Method

The proposed approach uses an impedance sensor developed by CTTM and LAUM [4]. This sensor performs accurate impedance measurements at low frequencies. Typically, the frequency range 20Hz – 5000Hz can be covered which offers new possibilities compared to multi-microphones duct measurements methods.

The porous density and bulk modulus are obtained from the measurement of two impedances (Figure 1). The surface impedance of the porous sample is measured for two mounting conditions:

- Z_s^{∞} is the measured surface impedance with a rigid back (infinite impedance),
- Z_s^A is the measured surface impedance with the sample loaded by the impedance Z_A.

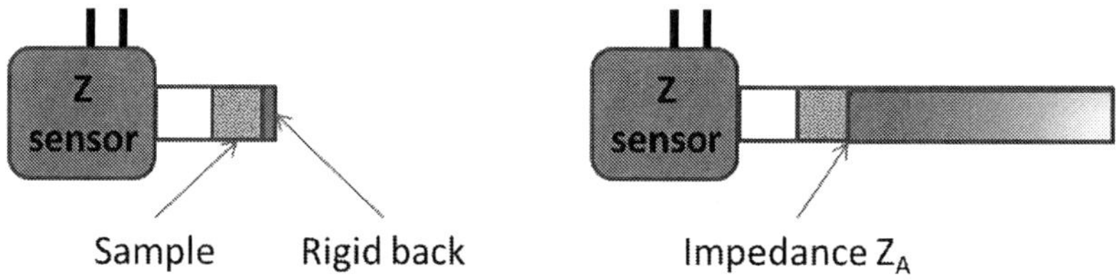

Figure 1: Measurement procedure with two acoustic loads

Then, the equivalent characteristic impedance and wave number of the porous media are given by:

$$Z_{eq}^2 = Z_s^A(Z_A + Z_s^{\infty}) - Z_A Z_s^{\infty}$$
$$k_{eq} = \arctan(-jZ_{eq}/Z_s^{\infty})/d$$

where d is the sample thickness. Dynamic density is calculated with $\rho_{eq} = Z_{eq}k_{eq}/\omega$ and bulk modulus with $K_{eq} = \omega Z_{eq}/k_{eq}$.

Considering the poroelastic media described by the JCAL model [5], the low frequency asymptotes of dynamic density and dynamic bulk modulus allow the estimation of the open porosity ϕ and the static air flow resistivity σ by:

- $\phi = P_0 / \lim_{\omega \to 0} \left(real(K_{eq}) \right)$
- $\sigma = -\omega \times \lim_{\omega \to 0} \left(imag(\rho_{eq}) \right)$

where P_0 is the atmospheric pressure.
To optimise the approach, it is better to take non resonant acoustic loads. Thus, for the impedance Z_A, we have chosen a high performance and compact anechoic end [6].

3 Results

Figure 2 below presents some results for the estimation of open porosity and static air flow resistivity, showing that measurement below 100Hz are often requested to reach asymptote value of the parameters.

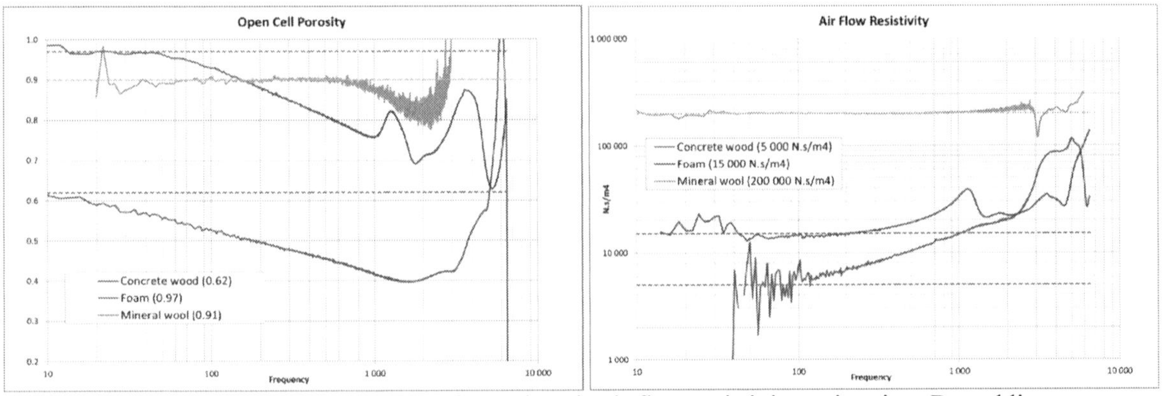

Figure 2: Example of open porosity and static air flow resistivity estimation. Dotted lines correspond to final parameter estimation.

In the presentation, we will show that the proposed method allows the analytical estimation of the complete set of poro-elastic fluid equivalent parameters with only one compact device. All parameters are estimated from the same sample, and with the same mounting condition, thus providing the method with increased accuracy.

4 References

[1] R. Panneton and X. Olny, "*Acoustical determination of the parameters governing viscous dissipation in porous media*" , J. Acoust. Soc. Am. 119. 2027-2040 (2006),
[2] X. Olny and R. Panneton, "*Acoustical determination of the parameters governing thermal dissipation in porous media*" , J. Acoust. Soc. Am. 123. 814-824 (2008),
[3] M. Niskanen, J.-P. Groby, A. Duclos, O. Dazel, J. C. Le Roux, N. Poulain, T. Huttunen, and T. Lahivaara, "*Deterministic and statistical characterization of rigid frame porous materials from impedance tube measurements*", J. Acoust. Soc. Am., **142**: 2407-2418, 2017.
[4] J.C. Le Roux, J.P. Dalmont,"A New Impedance Sensor for Industrial Applications", *Acoustics 2012, Nantes, France* (2012)
[5] D. Lafarge, P. Lemarinier, J.-F. Allard and V. Tarnov, "*Dynamic compressibility of air in porous structures at audible frequencies*", J. Acoust. Soc. Am. 102. 1995-2006 (1997),
[6] F. Fohr, E. Portier, "*Conduit comprenant au moins une terminaison anéchoique…*", French patent FR 1771310 (2017).

Deterministic and statistical methods for the characterisation of poroelastic media from sound absorption measurements

J. Cuenca[1], P. Göransson[2], L. De Ryck[1], T. Lähivaara[3]

[1] Siemens Industry Software, Interleuvenlaan 68, 3001 Leuven, Belgium
[2] Department of Aeronautical and Vehicle Engineering, KTH Royal Institute of Technology, Teknikringen 8, 10044 Stockholm, Sweden
[3] Department of Applied Physics, University of Eastern Finland, P.O. Box 1627, 70211 Kuopio, Finland

A framework is proposed for the characterisation of a rigid-backed poroelastic sample by using sound absorption measurements. The set of parameters of interest includes five transport parameters, two elastic properties, the mass density and the thickness of the sample. In the present work, all eight parameters are considered unknown and treated as random variables, whose nominal values and uncertainty ranges are of interest. The parameter estimation relies on model inversion, for which two complementary procedures are used, namely using deterministic and statistical tools. The methodology is applied to measurements in a two-microphone impedance tube.

1 Problem

The proposed approach aims at estimating the properties of poroelastic material samples from sound absorption measurements. The use of sound absorption coefficients provides experimental advantages with respect to measurements of the sound transmission behaviour. For instance, the measurement can be performed using a two-microphone impedance tube [1], and the rigid backing condition is relatively simple to set up. Despite the inherent simplicity and robustness of sound absorption measurements, these convey less information than the intrinsic acoustic properties of the sample such as the bulk modulus and equivalent density. The Biot-Johnson-Champoux-Allard model [2] for poroelastic media is here considered, which allows for the problem to be modelled analytically as a function of the eight unknowns.

2 Method

A recent method for the characterisation of dynamical systems [3, 4, 5] demonstrates the benefits of performing the model inversion in ascending frequency increments, as a series of sub-problems. This technique allows to gradually guide the parameter values towards the global solution of the problem, without terminating at local minima. In practice, each sub-problem is solved by means of a deterministic optimisation tool, such as the globally convergent method of moving asymptotes [6]. A statistical inversion approach is then used in order to refine the parameter estimates and obtain their uncertainty ranges. This is performed within the Bayesian framework using a Markov chain Monte Carlo approach [7].

3 Characterisation of a melamine foam sample

The present work applies the method to the experimental characterisation of poroelastic materials in a two-microphone impedance tube. Table 1 shows the obtained solution for a 24 mm-thick sample of melamine foam. The parameter estimates are provided as the maximum a posteriori estimate and their uncertainty ranges as the 90% credible intervals.

References

[1] ISO 10534-2. *Acoustics – Determination of the sound absorption coefficient and impedance in impedance tubes – Part 2: Transfer-function method*, 1998.

[2] J. Allard and N. Atalla. *Propagation of Sound in Porous Media: Modelling Sound Absorbing Materials 2e.* Wiley, 2009.

Parameter	Unit	Estimate	Uncertainty range
Thickness	mm	23.14	23.01 ... 23.29
Mass density	kg·m^{-3}	11.958	11.562 ... 12.449
Porosity	-	0.9996	0.9906 ... 0.9998
Flow resistivity	kN·s·m^{-4}	18.709	18.516 ... 19.044
Tortuosity	-	1.001	1.000 ... 1.023
Viscous characteristic length	μm	70.93	69.09 ... 74.82
Thermal characteristic length	μm	159.07	155.92 ... 163.24
Young's modulus	kPa	196.697	190.276 ... 203.756
Viscoelastic damping ratio	%	4.696	4.364 ... 4.982

Table 1: Parameter values and uncertainty ranges obtained from the proposed approach for a sample of melamine foam.

[3] P. Göransson, J. Cuenca, and T. Lähivaara. Parameter estimation in modelling frequency response of coupled systems using a stepwise approach. *Mechanical Systems and Signal Processing*, 126:161–175, 2019.

[4] J. Cuenca, P. Göransson, L. De Ryck, and T. Lähivaara. Inverse parameter estimation in resonant, coupled fluid-structure interaction problems. In *ISMA/USD*, Leuven, 2018/9/17-19.

[5] J. Cuenca, L. De Ryck, P. Göransson, and T. Lähivaara. Material parameter identification of coupled resonant systems using impedance tubes. In *26th International Congress on Sound and Vibration*, Montreal, 7-11 July 2019.

[6] K. Svanberg. A class of globally convergent optimization methods based on conservative convex separable approximations. *SIAM Journal on Optimization*, 12(2):555, 2002.

[7] J. Kaipio and E. Somersalo. *Statistical and Computational Inverse Problems*. Springer-Verlag, 2005.

Ultra-broadband Acoustic Barrier based on Hyper-dampened Fano resonance

Huy Nguyen[1], Sharon Tracy[2], Guoliang Huang[1]

[1] Department of Mechanical & Aerospace Engineering, University of Missouri, Columbia, MO 65211, United States.

[2] Materials Innovation, Steelcase Inc., Grand Rapids, Michigan 49508, United States of America

Ultra-broadband sound reduction schemes covering living and working noise spectra are of high scientific and industrial significance. Here we report, both theoretically and experimentally, on an ultra-broadband acoustic barrier assembled from space-coiling metamaterials (SCMs) supporting two Fano resonances. Moreover, acoustic hyper-damping is introduced by integrating additional thin viscous foam layers in the SCMs for optimizing the sound reduction performance. A simplified model is developed to study sound transmission behavior of the SCMs under a normal incidence, which sets forth the basis to understand working mechanism. An acoustic barrier with 220 mm thickness is then manufactured and tested to exhibit ultra-broadband transmission loss(TL) overall above 10 dB across the range 0.44−3.85kHz, covering completely nine third-octave bands. In addition, unconventional broadband absorption in the dampened barrier (65%) are experimentally observed as well. We believe this work paves the way for realizing effective broadband sound insulation, absorption and sound wave controlling devices with efficient ventilation.

Low-frequency broadband dissipation using Micro-Capillary Plates

T. Bravo[1], C. Maury[2]

[1] Instituto de Tecnologías Físicas y de la Información, Serrano 144, Spanish National Research Council, 28006 Madrid, Spain
[2] Aix Marseille Univ, CNRS, Centrale Marseille, 4 impasse Nikola Tesla, Laboratory of Mechanics and Acoustics (LMA), 13013 Marseille, France

1 Introduction

The study of new absorbers for control of low-frequency noise constitutes a continuous area of research. Micro-perforated panels (MPPs) [1] can be tailored to work optimally at low frequencies, but they are confined in a narrow frequency band as they build on the principle of Helmholtz resonances. To get rid of these limitations, the use of unbacked panels has been considered as an alternative to cavity-backed partitions. Optimization results have shown that reducing the size of the hole diameters down to 10 μm can provide wideband absorption, but at the expense of an increase of the manufacturing cost. This work aims at investigating the acoustic properties of ultra-perforated membranes, denoted micro-capillary plates (MCPs). Although these devices are currently used as image intensifiers or detectors of cosmic rays, they also found potential applications in the field of acoustics to achieve low-frequency anechoic terminations [2]. A MCP is a slab made up of a resistive material with typical thickness between 1 mm and 5 mm, composed of a regular array of slots or micro-channels densely distributed over the whole surface. The micro-channel diameters are between 10 μm and 50 μm and typical perforation ratios vary between 50% and 70%.
In this work, we further investigate the potential of MCPs as wideband acoustic absorbers with respect to other traditional absorbers as well as their optimization. An analytical effective approach will be presented that describes different working flow regimes. It will enable to perform efficient parametric studies that provide the performance prediction of MCPs as a function of their constitutive physical parameters. Results are compared for different MCPs, working in the slip-flow regime, against those corresponding to classical MPPs, operating in the continuum regime. The simulation results obtained for optimised MCP configurations will be validated by measurements of their absorption coefficient in a Kundt tube, showing good agreement with the analytical predictions. Moreover, we will carry out a comparison against published results using a wire-mesh or a textile screen as resistive layers that are able to provide a controlled resistance suited to the particular problem of interest.

2 Micro-Capillary Plate modelling

Rarefaction effects should be considered in the field of microfluidics when characteristics lengths are of the order of 1 μm [3], as in the case in MCPs micro-channels. Well-established continuum laws may then not be valid. In particular, a regime classification is done in terms of the ratio between the molecular mean free path λ and the characteristic length L of the control volume, that is taken for MCPs as the diameter of the perforations. This quantity is called the Knudsen number, $\mathrm{Kn} = \lambda / L$: it characterizes rarefaction effects within the micro-channels [3]. In particular, when $\mathrm{Kn} < 10^{-3}$, the flow is in the continuum regime and classical continuity boundary conditions at the channel walls are fulfilled. On the other hand, when $10^{-3} \leq \mathrm{Kn} \leq 10^{-1}$, the flow is in the slip-flow regime: a velocity and temperature jump have to be considered at the wall surface. Typically, most of the microsystems that work with gases are in the slip flow regime. Assuming straight cylindrical channels with longitudinal axis in harmonic regime ($\mathrm{e}^{\mathrm{j}\omega t}$), linearized momentum and energy conservation lead to the viscous transfer impedance of a micro-channel, given per unit length by

$$Z_v = \mathrm{j}\omega\rho_0 t\left\{1 - \frac{2}{k_v r_0}\frac{J_1(k_v r_0)}{\left[J_0(k_v r_0) - B_v k_v r_0 J_1(k_v r_0)\right]}\right\}^{-1} \tag{1}$$

where $k_v = \sqrt{-\mathrm{j}\omega\rho_0/\mu}$ is the viscous diffusion wavenumber, μ the air dynamic viscosity, ρ_0 the air density, and J_0 and J_1 are the Bessel functions of the first kind of orders 0 and 1 respectively. From Eq. (1), one can simulate the acoustic performances of MCP samples in slip-flow regime, which leads to an overall design chart of the MCP dissipation performance as a function of both the Knudsen number and the Shear number.

3 Experimental study

The validation of the simulated absorption results [4] is performed using a small Kundt tube with inner radius 1.5 cm. It is positioned vertically and connected on one side to a loudspeaker driven by a white noise signal up to 7 kHz. If unbacked, the MCP sample radiates on the other side into a semi-anechoic room. Measured absorption results are presented in Figure 1 in red for the unbacked and backed configurations. They are compared against predictions assuming different loads. As it can be appreciated, a good agreement is achieved for the unbacked configuration when assuming an unflanged radiation load. This load radiation impedance has been determined experimentally using a miniature pressure-velocity probe. Since the load behind the MCP significantly influences its acoustical performance, a discussion is provided of the sensitivity of the MCP dissipation performance to variations in the load impedance.

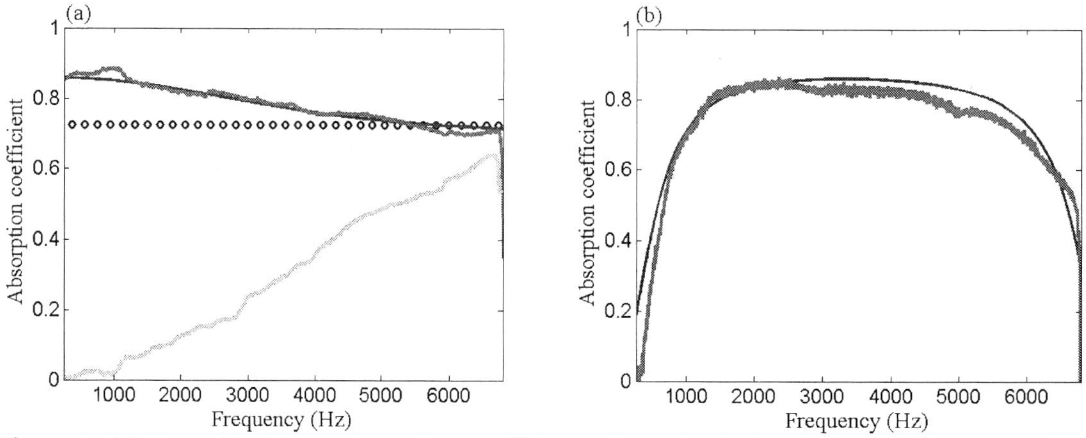

Figure 1 : (a) Normal incidence absorption coefficients of an unbacked optimized MCP: measured (red) and calculated assuming unflanged (blue) and anechoic (black circle) radiation loads, and that measured of an open termination without MCP (green); (b) Normal incidence absorption coefficient of an optimized MCP rigidly-backed by a cavity of depth 24 mm: measured (red) and calculated (blue).

4 Comparison with other absorbers performances

Many acoustic materials used in demanding environments in the fields of aeronautics or ventilating systems are covered with a resistive layer, initially used for protection, but whose parameters can be selected to maximize the normal sound absorption of such rigidly-backed porous layers [5]. The performance of MCPs backed by a downstream porous media will be compared against wire-mesh or textile screens in order to show their advantages and weaknesses in relation to other sound controlling devices.

References

[1] D.Y. Maa, *Potential of microperforated panel absorbers*, J. Acoust. Soc. Am. **104**, pp. 2861–2866, 1988.

[2] J.-P. Dalmont, J. Kergomard and X. Meynial, *Realization of an anechoic termination for sound ducts at low frequencies*, C. R. Acad. Sci. Paris **309** II, pp.453–458, 1989.

[3] S. Kandlikar, S. Garimella, D. Li, S. Colin and M. R. King, *Heat Transfer and Fluid Flow in Minichannels and Microchannels*, Elsevier Ltd., Oxford, 2nd Edition, 2014.

[4] C. Maury and T. Bravo, *Wideband sound absorption and transmission through micro-capillary plates: modeling and experimental validation*, J. Sound Vib. **478**, 115536 (2020).

[5] F. Chevillotte, *Controlling sound absorption by an upstream resistive layer*, Appl. Acoust. **73**, pp. 56-60, (2012).

Acoustical and non-acoustical behaviour of nanofibers membranes

K. V. Horoshenkov, A. Hurrell

Department of Mechanical Engineering, University of Sheffield, Sheffield, S1 3JD, UK

1 Introduction

There is a general lack of publications on the acoustical and related non-acoustical properties of nanofibrous media. This work attempts to contribute to this gap and to highlight problems associated with acoustic and related non-acoustic characterisation of these materials. It makes use of Biot- and Darcy-type mathematical models to explain the observed acoustical and related non-acoustical behaviours of the nanofibres. It identifies theoretical gaps related to the physical phenomena which can be responsible for the observed acoustical behaviours of nanofibrous membranes and it presents recommendations to fill these gaps. The novelty of this work is in the application of a robust theoretical model to explain the measured acoustical behaviour of thin nanofibrous membranes placed on a foam substrate. With this model the actual flow resistivity of nanofibers is estimated from acoustical data. It is demonstrated that a classical model for the flow resistivity of fibrous media does not work when the Knudsen number becomes greater than 0.02, i.e. then the diameter of nanofibres becomes comparable with the mean free path.

2 Material preparation

19 membrane samples were electrospun using in-house built single-needle electrospinning rigs at the University of Surrey and the University of Sheffield. Pump-rates were tailored to each solution, voltage, and atmospheric conditions, increasing the rate to the maximum possible before unwanted dripping from the spinneret was observed. This ensured a stable, steady stream of polymer for optimum nanofibre production in the fibre diameter range of 281 – 3630 nm and membrane thickness of 5.41 – 32.38 μm.

3 Material characterisation

The acoustical properties of the fibrous membranes were measured at the University of Sheffield in the 45 mm impedance tube in the presence of the melamine foam substrate. The 3-parameter model for the acoustical properties of porous media [1] was used to fit the measured complex acoustic reflection coefficient of melamine foam. The fibre diameter was measured using SEM images. The fibre density and porosity were measured directly from sample volume and mass information. The acoustical and related non-acoustical properties of nanofibrous media were inverted from the complex reflection coefficient data measured in the frequency range of 300 and 300 Hz.

4 Results

Figure 1 illustrates the effect of a 22 μm thick, 15 kV PMMA nanofibrous membrane on the complex reflection coefficient of the melamine foam substrate. The continuous lines correspond to the predictions by the model detailed in ref. [1] and markers correspond to the measured data. The presence of a very thin nanofibrous membrane causes a significant reduction in the real and imaginary parts of the reflection coefficient. This effect is explained by the increase in the real part of the surface impedance of the melamine foam substrate when the thin, nanofibrous layer is added on the top it.

According to the Kozeny-Carman model [2] the relation between the flow resistivity, σ, and fibre diameter, $\underline{d}$, is:

$$\frac{\sigma \underline{d}^2}{\eta} = \frac{180(1-\phi)^2}{\phi^3}. \quad (1)$$

Here η is the dynamic viscosity of air. The Kozeny-Carman model works well to predict the flow resistivity of highly porous media composed of fibres with the diameter of a few microns [2]. However, it is interesting to check if it also works for fibrous media made of nanofibres. To the best of our knowledge this has not been done before for flow velocities through the fibres which are comparable with the acoustic velocity in the audible range.

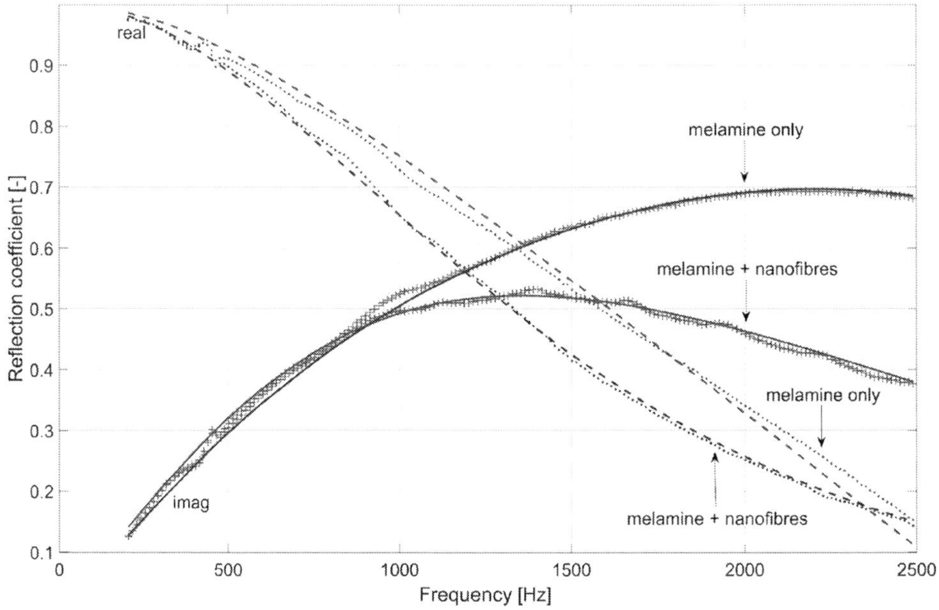

Figure 1 : The effect of nanofibrous membrane on the real and imaginary part of a 16 mm thick layer of melamine foam. Lines correspond to the theoretical fit, dots and crosses correspond to the measured data.

Let us denote the left- and right-hand parts of eq. (1) as:

$$f_1(\sigma, \underline{d}) = \frac{\sigma \underline{d}^2}{\eta}, \tag{2}$$

and

$$f_2(\phi) = \frac{180(1-\phi_{nf})^2}{\phi_{nf}^3}, \tag{3}$$

respectively. We can call these two functions the flow resistivity, $f_1(\sigma)$, and porosity, $f_2(\phi_{nf})$, terms, respectively.

The work by Umnova et al [3] suggests that the flow resistivity of in eqs. (2) and (3) needs to be compensated for the so-called no-slip effects when the fibres diameter, $\underline{d}$, becomes comparable tof mean free path, l. According to this work the flow resistivity of nanofibres compensated for no-slip condition is (eq. (46) in [3]:

$$\sigma_{\perp} = \frac{\sigma}{1 + \frac{4K_n}{1+2K_n} F(\phi_{nf})}, \tag{4}$$

where

$$F(\phi_{nf}) = \frac{\phi_{nf}^2}{-2\ln\ln(1-\phi_{nf}) - 2\phi_{nf} - \phi_{nf}^2} \qquad (5)$$

and σ is predicted by eq. (1). It is logical to assume that functions (2) and (3) should be equal if eq. (1) holds for nanofibres.

Let us now use eqs. (2)-(4) to check how eq. (1) holds for nanofibres. Figure 2 graphically illustrates the dependence of these functions on the Knudsen number, $K_n = \frac{l}{d}$, where l=68 nm is the mean free path in air. If eq. (1) were to hold true, then the two sets of data shown in Figure 2 would be close to the solid black line which corresponds to the value of $f_2(0.8)$; $\phi_{nf}=0.8$ is a typical value of porosity fibrous media made from relatively large fibres. According to Figure 2 this seems only the case for fibres with relatively large mean diameters, i.e. for $K_n \approx 0.02$. As the fibre diameter decreases, the difference between the values of $f_1(\sigma_{nf}, \underline{d})$ and $f_2(\phi_{nf})$ becomes greater. The flow resistivity of nanofibres decreases with an increase in K_n. On the other hand, the function $f_2(\phi_{nf})$ grows with an increase in K_n. This suggests that equation (1) no longer holds for $K_n > 0.02$ or for smaller fibre diameters. The compensation for nonslip conditions (eq. (4)) does not explain the drop in the flow resistivity of nanofibres with an increase in K_n (stars in Figure 2). The stars in Figure 2 show that the difference between the flow resistivity compensated for noslip conditions, $\sigma_{\perp}$, and measured flow resistivity, σ_{nf}, becomes more than one order of magnitude when the Knudsen number becomes greater than 0.1.

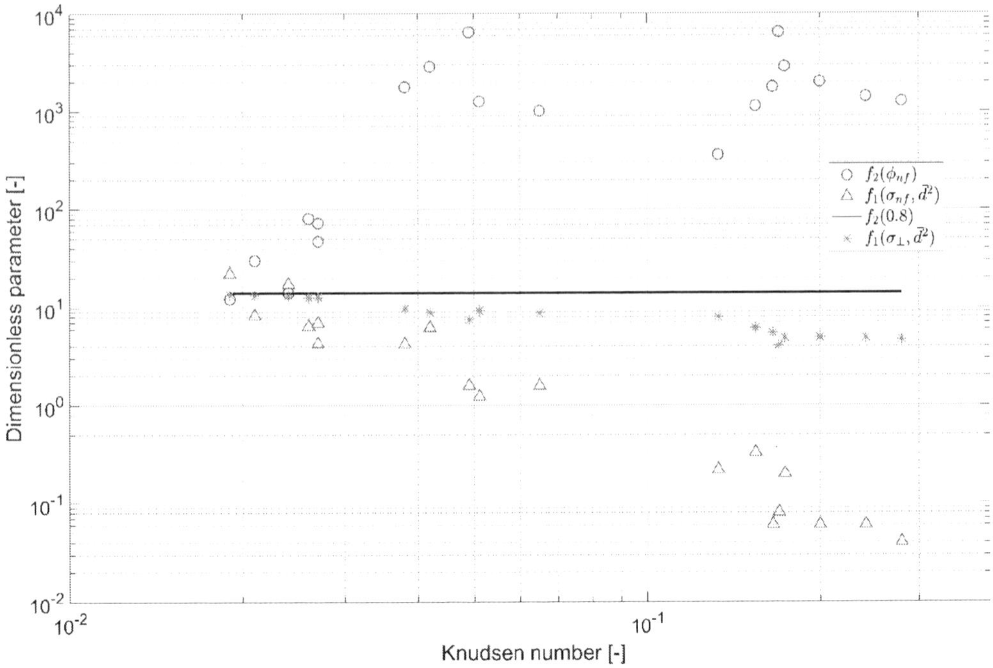

Figure 2 : The normalised flow resistivity and porosity terms in the Kozeny-Carman equations predicted for the parameters of nanofibrous media developed in this work.

References

[1] K. V. Horoshenkov, A. Hurrell, J-P Groby, *A three-parameter analytical model for the acoustical properties of porous media* J. Acoust. Soc. Am. **145**(4), pp. 2512-2517, 2019.

[2] P.C. Carman, *Fluid Flow through Granular Beds*, Trans. Instn Chem. Engrs., **15**, S32-47, 1937.

[3] O. Umnova, D. Tsiklauri, R. Venegas, *Effect of boundary slip on the acoustical properties of microfibrous materials* J. Acoust. Soc. Am. **126**(4), pp. 1850-1861, 2009.

Acoustic Characterisation of Thin Layer Nanofibers

V. Chen[1], I. Perez Pablos[2], J.M. Chen[3]

[1] Dyson Singapore Technology Centre, 2 Science Park Dr, #03-13/26 The Ascent, Singapore 118222
[2] Dyson Limited, Tetbury Hill, Malmesbury SN16 0RP, United Kingdom
[3] Singapore University of Technology and Design, 8 Somapah Rd, Singapore 487372

The acoustic absorption of porous media can be significantly improved by adding a nanofiber layer on the incident surface. Through acoustic characterisation of the nanofiber layer, a predictive model for nanofiber-porous systems was established. Subsequently, relating the nanofiber morphology to it's acoustic parameters allows the design of nanofiber-porous systems directly from fabrication parameters.

1 Introduction

The impact of nanofiber layers on acoustic absorption of a porous substrate has been demonstrated in [1, 2] and shown to be influenced by areal density [3]. Statistical optimisation of different configurations using multi-criteria decision analysis was investigated in [4]. Nanofibers as a bulk material were shown to be more absorptive than conventional porous media of equal thickness [5, 6] and acoustically characterised using the limp-frame model by [7] for predicting absorption coefficient. Artificial neural networks were also used to carry out similar predictions [8].
This work focuses on the acoustic characterisation of nanofibers as a thin layer to enable transfer matrix modelling of nanofibers over porous substrates, henceforth referred to as nanofiber-porous systems. Building on prior work on nanofibers [9] and resistive layers [10], the characterisation methodology for resistive screens in [11] was applied to nanofibers to inversely derive JCA [12] parameters and a lumped acoustic impedance. A relationship with nanofiber physical properties was then obtained using regression.

2 Sample Fabrication and Measurement

The electrospinning precursor was PA6 polymer solution at 5%, 10% and 15% concentration, giving nanofibers of different diameters measured using SEM imaging and DiameterJ [13]. Varying the electrospinning duration gave samples of different areal densities and layer thickness together with a standard PP non-woven fabric backing was measured using a Hanatek FT3 Precision Thickness Gauge as per ISO 9073-2:1995 with foot pressure of 0.5kPa. Static flow resistance was measured using a TexTest F3300 at the lowest possible flow velocity for each sample. Static flow resistivity σ, porosity ϕ and areal density ρ_A were computed from these measurements for comparison with acoustically-characterised values.

3 Characterisation Methods

The nanofiber samples were framed using sheet Mylar for consistency of mounting in the impedance tube over an air cavity. Based on simplifying assumptions of identical, parallel cylindrical perforations from [14], the nanofiber dynamic mass density $\tilde{\rho}_{\mathrm{c,nf}}$ was computed from surface impedance measurements using Eqn 13 from [11], which also details the derivation of JCA parameters to give characteristic acoustic impedance Z_c and wave number k_c. A lumped acoustic impedance Z_L was also computed by subtracting the analytical solution for surface impedance of the air gap from the same measurements. The melamine foam substrate was characterised using the two thickness method [15].

4 Results and Discussion

The characterised JCA and lumped element parameters were used in transfer matrices to model nanofiber-porous systems and predict absorption coefficient for comparison with measurement. Higher polymer concentration gave larger fiber diameters and pore sizes, resulting in lower flow resistance for the same areal density. The reduction in absorption coefficient was not proportional with change in flow resistance and this is still being studied. Nevertheless, within each batch of the same concentration, increasing areal density via longer electrospinning duration gave proportionately higher flow resistance and absorption coefficient. Results for the best performing batch are shown in Figure 1.

4.1 Prediction vs Measurement

Strong agreement in absorption coefficient was observed using both the JCA and lumped element models for samples A1-A5 in nanofiber-porous systems while in sample A6 the peak frequency was slightly over-predicted. Hypothesised causes were vibrational waves in the nanofiber layer or flow distortion at the nanofiber-porous interface. Biot theory [16] was used to account for the former and for the latter, dynamic tortuosity correction [14] was applied to the JCA parameters and an empirical offset added to $\mathrm{Im}(Z_L)$. This offset was based on the $(1 + \delta/\Lambda)$ factor observed in the high frequency expression of mass density in Equation 4 of [14] comprising the boundary layer thickness δ and viscous characteristic length Λ. Predictions with these 3 approaches are shown together with measurement in Figure 1; the $(1 + \delta/\Lambda)$ offset gave the best agreement and will be the subject of further study.

4.2 Physical vs Acoustic Characterisation

Directly-measured σ and ϕ exhibited some correlation with acoustically-characterised values within each batch of nanofibers, but differed by several orders of magnitude as similarly observed in [17]. In this context, the JCA parameters derived via [11] serve as effective values for modelling purposes. This disparity is another area of investigation, possibly via other methods of obtaining JCA parameters such as in [18].

To establish the feasibility of predicting nanofiber acoustic performance directly from physical parameters, Z_L was simplified into the constants $\mathrm{Re}(Z_L)$ and $\Delta\mathrm{Im}(Z_L)$ for regression model outputs. Applying iterative forward selection to physical parameters as regression model inputs, ρ_A was identified as the primary predictor for $\mathrm{Re}(Z_L)$ and $\Delta\mathrm{Im}(Z_L)$ while the latter output also had fiber diameter as a secondary predictor, giving R^2 of 0.8 to 0.9 at 95% confidence interval.

This regression model is the first step towards a design tool for specifying acoustic performance directly from electrospinning parameters. The application of the $(1 + \delta/\Lambda)$ offset to Z_L predicted from this regression without *a priori* characterisation of Λ will also be further investigated, alongside more detailed regression modelling using JCA parameters as outputs.

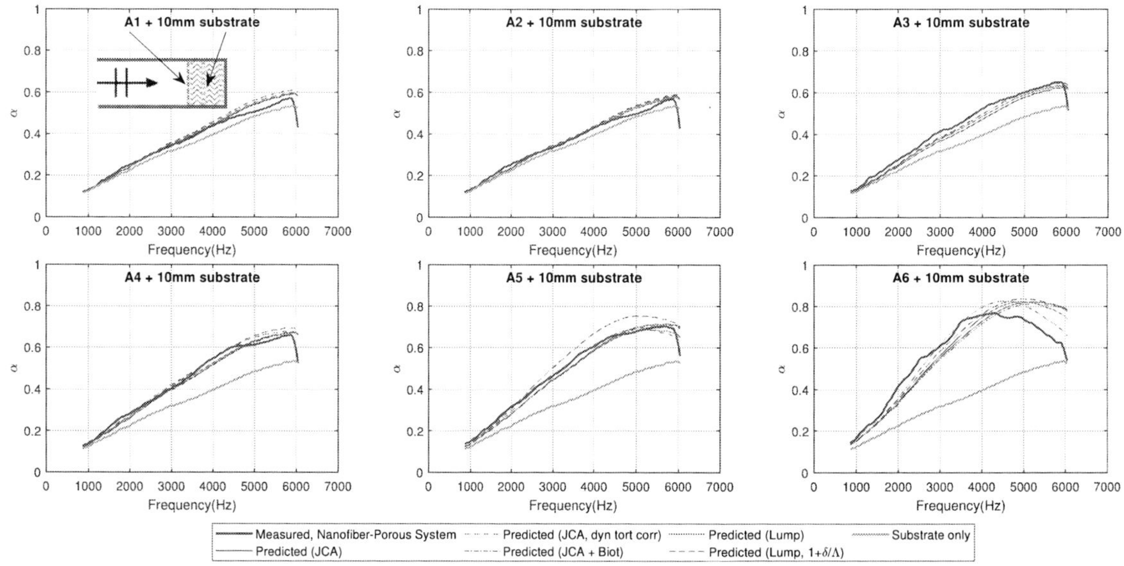

Figure 1: Predicted and measured absorption coefficients for nanofiber-porous systems (5% conc, 10mm thick substrate) with corrections applied. A1 to A6 are in increasing order of areal density.

References

[1] Hai-fan Xiang, Shuai-xia Tan, Xiao-lan Yu, Yu-hua Long, Xiao-li Zhang, Ning Zhao, and Jian Xu. Sound absorption behavior of electrospun polyacrylonitrile nanofibrous membranes. *Chinese Journal of Polymer Science*, 29(6):650–657, aug 2011.

[2] Amir Rabbi, Hossein Bahrambeygi, Ahmad Mousavi Shoushtari, and Komeil Nasouri. Incorporation of nanofiber layers in nonwoven materials for improving their acoustic properties. volume 8, page 155892501300800. SAGE Publications, dec 2013.

[3] Ayşe Özkal and Funda Cengiz Çallıoğlu. Effect of nanofiber spinning duration on the sound absorption capacity of nonwovens produced from recycled polyethylene terephthalate fibers. *Applied Acoustics*, 169:107468, dec 2020.

[4] M. Kucukali-Ozturk, B. Nergis, and C. Candan. Design of layered structure with nanofibrous resonant membrane for acoustic applications. *Journal of Industrial Textiles*, 47(7):1739–1756, may 2018.

[5] Ws Khan, R. Asmatulu, and M. B. Yildirim. Acoustical properties of electrospun fibers for aircraft interior noise reduction. *J. Aerosp. Eng.*, 25(3):376–382, 2012.

[6] Youngjoo Na, Tove Agnhage, and Gilsoo Cho. Sound absorption of multiple layers of nanofiber webs and the comparison of measuring methods for sound absorption coefficients. *Fibers and Polymers*, 13(10):1348–1352, dec 2012.

[7] Shuichi Akasaka, Takahisa Kato, Keisuke Azuma, Yuichi Konosu, Hidetoshi Matsumoto, and Shigeo Asai. Structure-sound absorption property relationships of electrospun thin silica fiber sheets: Quantitative analysis based on acoustic models. *Applied Acoustics*, 152:13–20, sep 2019.

[8] Giuseppe Ciaburro, Gino Iannace, Jessica Passaro, Aurelio Bifulco, Daniele Marano, Michele Guida, Francesco Marulo, and Francesco Branda. Artificial neural network-based models for predicting the sound absorption coefficient of electrospun poly(vinyl pyrrolidone)/silica composite. *Applied Acoustics*, 169:107472, dec 2020.

[9] Kirill Horoshenkov, Alistair Hurrell, Mohan Jiao, and Michael Pelegrinis. What is the actual influence of a nano-fibrous membrane on the acoustical property of porous substrate? 2 nanofibrous membranes 2.1 structure. volume 144, pages 1755–1755. Acoustical Society of America (ASA), 04 2018.

[10] F. Chevillotte. Controlling sound absorption by an upstream resistive layer. *Applied Acoustics*, 73(1):56–60, jan 2011.

[11] L. Jaouen and F. X. Bécot. Acoustical characterization of perforated facings. *The Journal of the Acoustical Society of America*, 129(3):1400–1406, 2011/03/01 2011.

[12] Yvan Champoux and Jean-F. Allard. Dynamic tortuosity and bulk modulus in air-saturated porous media. *Journal of Applied Physics*, 70(4):1975–1979, aug 1991.

[13] Nathan A. Hotaling, Kapil Bharti, Haydn Kriel, and Carl G. Simon. Diameterj: A validated open source nanofiber diameter measurement tool. *Biomaterials*, 61:327–338, 2015.

[14] Noureddine Atalla and Franck Sgard. Modeling of perforated plates and screens using rigid frame porous models. *Journal of Sound and Vibration*, 303(1):195 – 208, 2007.

[15] C. D. Smith and T. L. Parrott. Comparison of three methods for measuring acoustic properties of bulk materials. *Journal of the Acoustical Society of America*, 74(5):1577–1582, 1983.

[16] Raymond Panneton. Comments on the limp frame equivalent fluid model for porous media. *The Journal of the Acoustical Society of America*, 122(6):EL217–EL222, dec 2007.

[17] Kirill Horoshenkov, Alistair Hurrell, and Mohan Jiao. Are nano-fibers an emerging noise control solution? *The Journal of the Acoustical Society of America*, 144(3):1755–1755, sep 2018.

[18] Kirill V. Horoshenkov, Alistair Hurrell, and Jean-Philippe Groby. A three-parameter analytical model for the acoustical properties of porous media. *The Journal of the Acoustical Society of America*, 145(4):2512–2517, apr 2019.

Acoustical properties of TEOS based granular silica aerogels

H. Begum[1], K. V. Horoshenkov[1], M. Conte[2], W. J. Malfait[3], S. Zhao[3], M. M. Koebel[3], P. Bonfiglio[4], R. Venegas[5]

[1]*Department of Mechanical Engineering, The University of Sheffield, S1 3JD, United Kingdom*
[2]*Department of Chemistry, The University of Sheffield, S3 7HF, United Kingdom*
[3]*Laboratory for Building Energy Materials and Components, Swiss Federal Laboratories for Materials Science and Technology, Empa, Überlandstrasse 129, 8600, Dübendorf, Switzerland*
[4] *Materiacustica srl, Via C. Ravera 15A, 44122 Ferrara, Italy*
[5] *University Austral of Chile, Institute of Acoustics, P.O. Box 567, Valdivia, Chile*

Currently available data suggests that granulated aerogels can be of interest in terms of their sound absorption performance in the audio frequency range. However, there is still no thorough understanding of the complex physical phenomena which are responsible for their observed acoustical properties.

This paper makes use of advanced material characterization methods and a mathematical model to explain and predict the acoustical properties of granular silica aerogels. The samples are produced through a two-step, acid-base sol-gel process, with sol silica concentration and density being the main variables. The pore structure is carefully characterised by nitrogen sorption analysis and scanning electron microscopy, and the acoustical properties of hard-backed granular silica aerogels are measured in a bespoke acoustic impedance tube and the results predicted accurately with the adopted theoretical model. Although silica aerogels have over 90% porosity, this was not reflected in the parameter values predicted with the model. Results found that only a proportion of the micro and mesopores in the direct vicinity of the grain surface influenced the acoustical properties of aerogel. Further work in the hierarchical pore structure of aerogels is required to understand better the roles of different pore scales on the measured acoustical properties of granulated aerogel.

In this extended abstract we focus on one type of silica aerogel prepared with variable $PEDS_{P750}$ content in the sol to produce an aerogel named PEDS E30. Fig.1 shows that the particles are not spherical but angular with sharp edges. The complex reflection coefficient of these materials was measured in a specially developed impedance tube with 10 mm diameter in the frequency range of 300 – 3000 Hz.
Granular aerogels consist of highly porous particles with a large internal pore surface area. Modelling of their acoustical properties requires accounting for its multiscale nature and physical processes that occur at different scales. In this work, the model proposed by Venegas et al. [1] was applied. This upscaled analytical model has been developed for an array of spherical porous grains in which two inner-particle scales of porosity are considered. The model accounts for the viscosity and heat transfer effects in the voids formed between the particles, rarefied gas flow and heat transfer in the inner-particle transport pores, interscale (voids to/from inner-particle pores) pressure diffusion, interscale (transport- to/from mesopores) mass diffusion, and sorption in the micro- and mesopores.

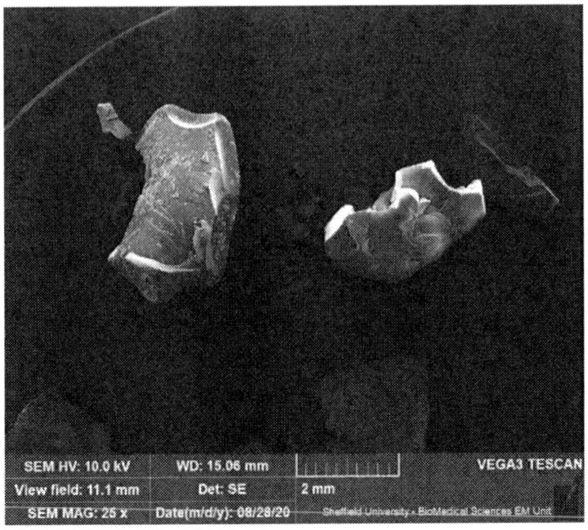

Figure. 1. SEM image of PEDS E30 showing micrometric grains of the 2-3 mm fraction.

After inserting the key experimental parameters obtained from this study into the model, Fig. 2 shows there is agreement between the measured and predicted acoustical surface impedance spectra. The relative mean error between the data and predictions was generally less than 3.1%, this suggests that the model captures the acoustical behaviour of aerogel PEDS E30 accurately.

Figure 2. Measured and predicted reflection coefficient for a hard-backed layer of aerogel granulate PEDS E30.

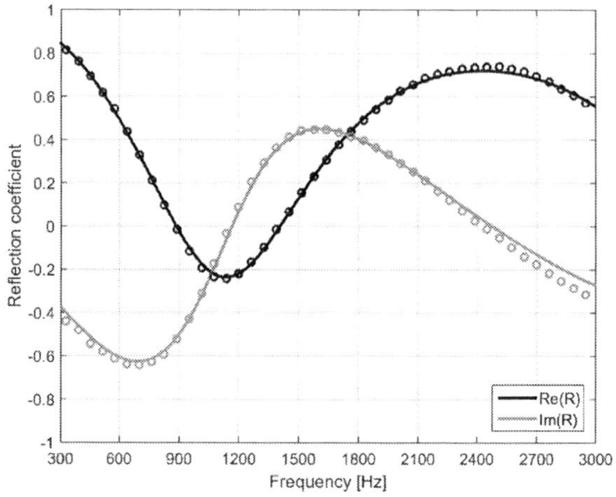

The results found the values of the inverted parameters to make physical sense. Silica aerogels have close to 100% open porosity, however this is not reflected in the parameter values inverted using the proposed mathematical model. Our findings indicate that not all the aerogel's mesopores appear to contribute to the observed acoustical performance.

Table 1. Measured properties and inverted parameters of PEDS E30.

Properties of PEDS E30	Measured	Inverted
r_p [mm]	1.50	0.85
d [mm]	50	52.9
ϕ [%]	93.5	57.7
mesopore radius [nm]	22.7	17.7

The radius, r_p of the macropores is in the order of 1 μm, which is a typical value for this type of transport pores (see ref. [2] for the case of activated carbon). The mesopore radius, measured from the SEM images, for PEDS E30 was 22.7 nm. Making use of the Knudsen diffusion coefficient, D_e the model inverts the nanopore radius parameter as 17.7 nm for PEDS E30. This value is close to those measured from the SEM images. The overall porosity, ϕ estimated from the porosity data is 57.7 whereas from the measured BET data was 93.5.

Aerogels usually consist of fully interconnected pores and this is not reflected in the pore parameter values inverted using the acoustical data and adopted sound propagation model. It is likely that only a proportion of the mesopore length which is in the direct vicinity of the transport pores or grain surface may influence the acoustical properties of the produced aerogels. This remains an open question and naturally suggests that more research is needed to understand better the relative roles of macro-, meso- and micropores on the acoustical properties of aerogels.

References

[1] Venegas, R. Boutin, C. Umnova, O., Acoustics of multiscale sorptive porous materials, Phys. Fluids, 2017;29(8):082006.
[2] Venegas, R. Umnova, O. Influence of sorption on sound propagation in granular activated carbon, J. Acoust. Soc. Am., 2016;140(2):755–766.

Acoustical investigation of aerogel granules modeled as a layer of poro-elastic material

Y. Xue[1], A. Dasyam[2], J. S. Bolton[3], B. Sharma[2]

[1]Seagate US LLC, Shakopee, MN 55379, USA
[2]Mechanics and Dynamics Laboratory, 1845 Fairmount St., Wichita State University, Wichita, KS, 67260
[3]Ray W. Herrick Laboratories, 177 S. Russell St., Purdue University, West Lafayette, IN, USA

1 Introduction

Aerogels are defined as mesoporous materials obtained by replacing the liquid phase within a gel by a gaseous phase, typically air. This underlying mesoporous structure provides aerogels unique macrostructural properties such as ultralow density, high transparency, and low thermal conductivity [1]. Given their ultralow density, aerogels are also an attractive, lightweight solution for noise control applications. Recent studies have shown that the acoustical properties of granular aerogels are very different from those observed for other granular media [2]. In this presentation, we present results from our recent investigation of the acoustical behavior of Enova IC3100 aerogel granules. Manufactured and commercially sold by Cabot Corporation, IC3100 aerogels are characterized by their comparatively smaller granule dimensions. Our previous experimental measurements [3] show that the acoustical performance of the IC 3100 aerogel granules differs from conventionally used sound absorption materials; multiple, lightly damped depth resonances with large peak values of absorption coefficients are observed low frequencies. Here, we present results from our attempt to model the acoustical behavior of IC3100 aerogels. The acoustical-related bulk properties required for the Johnson-Champoux-Allard (JCA) model are calculated using an inverse characterization approach. These properties are then used to model the acoustical behavior of the granular aerogel layer using the Biot theory for porous media.

2 Methods

The aerogel's bulk properties were first input into the Johnson-Champoux-Allard (JCA) model [4,5] for the prediction of the material's bulk modulus and effective density. The Biot theory [6] was then applied to predict the fundamental acoustical properties such as the material's wavenumber(s) based on the bulk modulus and effective density predicted by using the JCA model. The characterization of the bulk properties was achieved by fitting the experimentally measured normal incidence absorption coefficients of a layer of aerogel resting on a hard-backing using a JCA-Biot-Transfer Matrix Method/Arbitrary Coefficient Method formulation [7,8], and by identifying the bulk properties which would provide the closest modeled results to the measured data. We examine the choice of modelling the aerogels as either a limp porous medium or as a poro-elastic medium.

3 Results

To avoid additional complexity, the IC 3100 aerogel was originally assumed to be limp porous medium, and the characterized bulk properties obtained by using Biot limp porous model (limp model) are shown in the second row of Table 1. On the other hand, the absorption coefficients of 1-inch and 2-inch layers of IC 3100 aerogel granules modeled given the limp model-characterized bulk properties are shown as the green-dot-dashed lines in Figure 1 (a) and (b), respectively. By comparing the green lines with the measured results (red solid lines), an obvious under-estimation of absorption coefficients can be observed in the low frequency area, which indicates an under-estimation of damping at this region. Therefore, to take one step forward, the aerogel was then modeled as a poro-elastic medium by using the Biot poro-elastic model (elastic model) to take the material's bulk elasticity properties including the loss factor, η_m, into account. There were 7041 line-spaced frequencies ranging from 600 to 6100 Hz involved into the modeling process to match the measurement frequency steps setting. Correspondingly, by introducing a 7041-step loss factor that dynamically decreases as the frequency increases, and by making it logarithmically spaced in ranges as listed at the end of the second and third rows in Table 1 for the 1-inch and 2-inch cases, respectively, it can be observed by comparing the blue-dashed lines in Figure 1 with the measured data that the dynamic loss factor realizes better absorption fittings for both cases.

Table1: Bulk properties characterized by fitting the measurements.

IC 3100 aerogel granules	σ [Rayls/m MKS]	ϕ	α_∞	Λ [μm]	Λ' [μm]	ρ_b [kg/m^3]	E_1 [Pa]	ν	η_m
Limp model 1-inch / 2-inch	10,457,000	0.999	3	36.08	36.08	39.86	N/A	N/A	N/A
Elastic model 1-inch	10,457000	0.999	3	36.08	36.08	39.86	775	0.396	20 to 0.0063
Elastic model 2-inch	10457000	0.999	3	36.08	36.08	39.86	775	0.396	10 to 0.0001

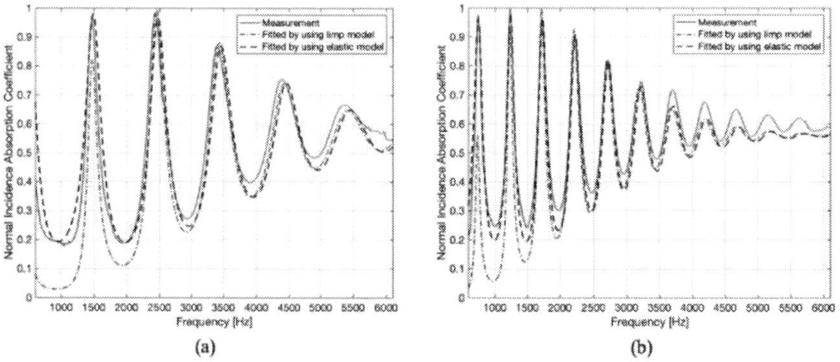

Figure 1: Absorption coefficients of (a) 1-inch, (b) 2-inch layer of IC 3100 aerogel granules.

4 Conclusions

The acoustical behavior of Enova IC3100 aerogels is modeled using the Biot theory for porous materials. Our results show that IC3100 granules show unique low-frequency acoustical properties that cannot be adequately captured using Biot's limp porous media formulation. Instead, we show that the poro-elastic model with a logarithmically decreasing loss factor provides a better match with experimental data.

References

[1] Pierre AC. History of aerogels. Aerogels Handbook: Springer; 2011. p. 3-18.

[2] Malakooti S., Churu H., Lee A., Xu T., Luo H., Xiang N., et al, "Sound insulation properties in low-density, mechanically strong and ductile nanoporous polyurea aerogels," Journal of Non-Crystalline Solids. 2017;476:36-45.

[3] Dasyam A., Khaleel O., Sharma B., "Acoustic absorption properties of granular silica aerogels," In INTER-NOISE and NOISE-CON Congress and Conference Proceedings 2020 Oct 12 (Vol. 261, No. 6, pp. 48-58). Institute of Noise Control Engineering.

[4] Johnson D., Koplik J., and Dashen R., "Theory of dynamic permeability and tortuosity in fluid-saturated porous media," *J. Fluid Mech.* **176**, 379–402 (1987).

[5] Champoux Y. and Allard J.F., "Dynamic tortuosity and bulk modulus in air-saturated porous media," *J. Appl. Phys.* **70(4)**, 1975–1979 (1991).

[6] Biot M., "Theory of propagation of elastic waves in a fluid-saturated porous solid," *J. Acoust. Soc. Am.* **28(2)**, 168–191 (1956).

[7] Xue Y., Bolton J. and Liu Y., "Modeling and coupling of acoustical layered systems that consist of elements having different transfer matrix dimensions," *J. Appl. Phys.* **126**, 165012 (2019).

[8] Xue Y., "Modeling and design methodologies for sound absorbing porous materials when used as layered vibration dampers", *Ph.D. dissertation*, Purdue University (2019).

A Poro-Elastic Model for Activated Carbon Stacks

Zhuang Mo[1], Tongyang Shi[2], Seungkyu Lee[2], Yongbeom Seo[2], J. Stuart Bolton[1]

[1]Ray W. Herrick Laboratories, 177 S. Russell St.,
Purdue University, West Lafayette, IN, USA
[2]3M Company, 3M Center, St. Paul, MN, USA

1 Introduction

Porous materials are widely applied in the field of noise control owing to their ability to attenuate transmission and/or reflection of sound. Among these materials, activated carbon is of increasing interest due to its excellent absorption performance at low frequencies. In 2016, a rigid, triple porosity model for granular activated carbon (GAC) was proposed by Venegas and Umnova [1]. In the proposed model, a sorption process was introduced to account for the effect of the smallest scale pores, which is correlated with the high absorption at low frequencies [1,2]. To extend that model to account for the finite frame stiffness of the particle stack, a poro-elastic model for a stack of granular activated carbon is introduced here. The new model combines aspects of the Biot theory [4,5,6] and the rigid GAC model [1,2,3]. The stable computational approach proposed by Dazel et al. [7] was followed when implementing the poro-elastic model. A fitting procedure based on particle swarm optimization (PSO) was applied to identify the parameters of proposed model that then yield predictions closely matching measurement of the absorption coefficient of rigid-backed activated carbon stacks.

2 Proposed poro-elastic model

The GAC model is based on the assumption that the particles are spherical, and that the pores within the particles are cylindrical [1,2]. In the proposed model, the volume change of the inter-particle interstices caused by the frame elasticity is considered. Therefore, the fluid phase bulk modulus and equivalent density were obtained from the GAC model, while the macro porosity, ϕ_p, in the GAC model, was treated as the porosity of the whole stack.

3 Parameter fitting

The parameters of the proposed model were obtained by fitting the model prediction to measured absorption coefficients. The fitting procedure was implemented *via* the PSO algorithm, which was realized by using the toolbox 'Constrained Particle Swarm Optimization' [8]. Since the bulk density for the activated carbon particles can be measured, that information was utilized as a nonlinear constraint applied to the porosities. In the cases shown in the next section, the bulk density inferred from fitted porosities was constrained to lie within $\pm 5\%$ of the measured value. Two arrays of weightings applied to different frequencies were defined for the measurements obtained using a small tube (diameter 2.9 cm) and medium tube (diameter 6.35 cm), respectively. The weightings at very low frequencies were set to be zero to discard the unreliable measurement in that range. The weightings around the frequency where a quarter-wave frame resonance peak appears were set to 2, while the weighting at all other frequencies was set to 1. The intention here was to help ensure that the fitting procedure yielded results that accurately captured the frame resonance feature Therefore, the optimization problem can be expressed as,

$$\begin{aligned} \min_{\mathbf{x}} \quad & f(\boldsymbol{x}) = \mathbf{w}^{\mathrm{T}}(\mathbf{a} - \mathbf{a}_m)^2/N \\ \text{s.t.} \quad & \mathbf{x}_{lb} < \mathbf{x} < \mathbf{x}_{ub} \\ & 0.95\rho_m < \rho_c(1-\phi_{tb}) < 1.05\rho_m \end{aligned} \qquad [1]$$

where N denotes the number of measured data points; $\mathbf{x}$ represents the parameter vector of the poro-elastic model, which consists of frame stiffness, Poisson's ratio, loss factor, macroscopic porosity, mesoscopic porosity, microscopic porosity, particle radius, mesopore radius, micropore radius, Langmuir constant, and configurational diffusivity, with $\mathbf{x} = (E, \nu, \eta, \phi_p, \phi_m, \phi_n, r_p, r_m, r_n, b, D_c)$; $\mathbf{x}_{lb}$ and $\mathbf{x}_{ub}$ denote the lower and upper bound of the parameters; **a** and $\mathbf{a}_m$ denote predicted and measured absorption coefficient, respectively; and **w** denotes the weighting vector applied on the measurements.

4 Fitting results

The fitting results for 30mm-thick stacks of two types of activated carbon are shown in the figure below.

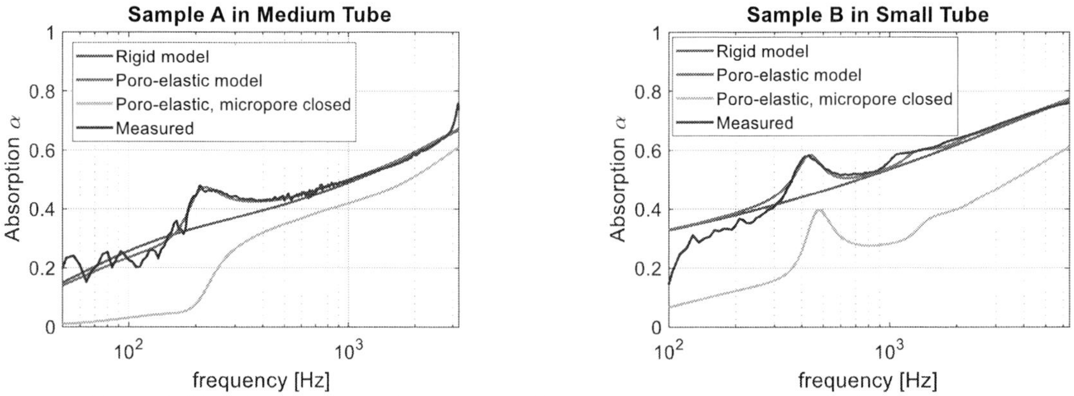

Figure 1: The absorption coefficient from fitted parameters compared with measurements.

In both cases, the fitted parameters capture the resonance peak at the frequency where the stack thickness equals one-quarter of the longitudinal structural wavelength. The predictions of the rigid model do not show that peak. In the second case, the fitted parameters can also capture the small second peak at the frequency where the stack thickness equals three-quarters of the longitudinal structural wavelength. Further, when the micropores are "turned off", the absorption is reduced, as shown by the yellow line, thus demonstrating the low frequency impact of the sorption process.

5 Conclusion

The proposed poro-elastic model can predict the high frequency asymptotic behavior of the activated carbon stack as well as the rigid model, but in addition, it can also reproduce the frame resonance peak(s) in the absorption coefficients, with physically reasonable parameters. When comparing the absorption coefficient with micropores open and closed, it can be seen that the existence of micropores boosts the absorption significantly.

References

[1] R. Venegas and O. Umnova, *Influence of sorption on sound propagation in granular activated carbon*, J. Acoust. Soc. Am. **140**, pp. 755-766, 2016.

[2] R. Venegas, C. Botin, and O.Umnova, *Acoustics of multiscale sorptive porous materials*, Phys. Fluids **29**, pp. 082006, 2017.

[3] T. Mellow, O. Umnova, K. Drossos, K. Holland, A. Flewitt, and L. Kärkkäinen, *On the adsorption-desorption relaxation time of carbon in very narrow ducts*, J. Acoust. Soc. Am. **123**, pp. 3141, 2016.

[4] O. Dazel, B. Brouard, C. Depollier, and S. Griffiths, *An alternative Biot's displacement formulation for porous materials*, J. Acoust. Soc. Am. **121**, pp. 3509-3516, 2007.

[5] J. F. Allard and N. Atalla, *Propagation of sound in porous media: modelling sound absorbing materials, second edition*, John Wiley & Sons, 2009.

[6] Y. Xue, *Modeling and design methodologies for sound absorbing porous materials when used as layered vibration dampers*, Purdue University PhD Thesis, 2019. https://doi.org/10.25394/PGS.9985331.v1.

[7] O. Dazel, J.-P. Groby, B. Brouard, and C. Potel, *A stable method to model the acoustic response of multilayered structures*, J. Appl. Phys. **113**, pp. 083506, 2013.

[8] S. Chen, *Constrained Particle Swarm Optimization (2009-2018)*, MATLAB File Exchange. https://www.mathworks.com/matlabcentral/fileexchange/25986.

High order X-FEM for the acoustic analysis of sound absorbing poro-elastic material with coupling interfaces

S.Wu[1], O. Dazel[2], G. Gabard[2], G.Legrain[1]

[1] École Centrale de Nantes, 1 rue de la Noë, 44321 Nantes, France
[2]Le Mans Université, 72085 Le Mans Cedex, France

Our work proposes a computational model based on the eXtended Finite Element Method (X-FEM) to predict the acoustic behaviour of porous structures coupled to other media. The poro-elastic material is modelled by a mixed formulation based on the Biot's thoery. This work demonstrates that the proposed method is able to alleviate the geometry limitation and to ensure the convergence of the Biot's equations simultaneously.

1 Background and problematic

Poro-elastic materials (PEMs) are widely used to build sound absorbers for practical noise control problem such as room acoustics ceiling and sound proofing in aircraft, cars and trains. Biot's thoery is one of common models to describe the energy absorption of the porous materials. In this work, we use a mixed formulation [1] based on the Biot's thoery as the model of the porous material. This formulation provides a full coupled description for the propagation of elastic solid and poro fluid waves, providing a natural coupling with other media.
Finite element methods (FEM) and wave based methods are two frequently used numerical methods for solving the Biot equations. The classical FEM suffers from convergence problems with respect to the number of degrees of freedom [2]. Wave based methods [3] show a better capacity to capture oscillation behaviour with less numbers of degrees of freedom. However, for most of wave based methods, extra operators at the interface are needed to couple all subdomains with different materials. Within the classical FEM, the mesh is needed to coincide with the geometry of the interface. If the mesh is not conforming at the interface, Lagrangian multiplier, penalty method and Nitsche's formulation are employed to weakly enforce the discontinuities between meshes as in wave based methods. All these requirements increase the computational cost especially when the methods are applied for complex geometry problems.

2 Proposed Method

To alleviate the geometrical limitation, the eXtended Finite Element Method (X-FEM) is taken into consideration. X-FEM was firstly proposed to solve crack growth problems without re-meshing [4] and was quickly applied to deal with material interfaces. In this work, we use a level-set function to locate the material interface implicitly. The elements split by the interface are enriched by a so-called ridge function [5] to introduce the weak discontinuity in the approximation inside the elements. Based on the idea of the X-FEM and the nature of the Biot equation, we provide the discretization strategies for two common coupling interfaces: acoustic-PEM and PEM-PEM. The Fields are approximated by high order Bernstein shape functions to ensure a proper rate of convergence for the Biot equation in high frequency.

3 Results and discussion

The sound pressure distribution in a car computed by the proposed method is given here. The driver seat is filled with XFM porous material modelled by the Biot formulation. An arbitrary normal velocity is imposed at the front windscreen under the frequency of 2000 Hz. A common situation where the driver's seat is switched from one position to another is simulated, which can be considered as a moving interface problem. The geometry of the problem and the mesh within our method are shown Figure 1. The profile of the driver seat is defined implicitly by level-set functions (in blue and red). Approximation of degree $p = 4$ is adopted to reach a converged solution using this mesh.
As seen in Figure 1, the discretization (meshes) is independent of the seat geometry as well as the position. The results in Figure 2 show that the variation of the acoustic pressure as the change of the seat position is well simulated and the absorption is captured in the region of the driver seat. This

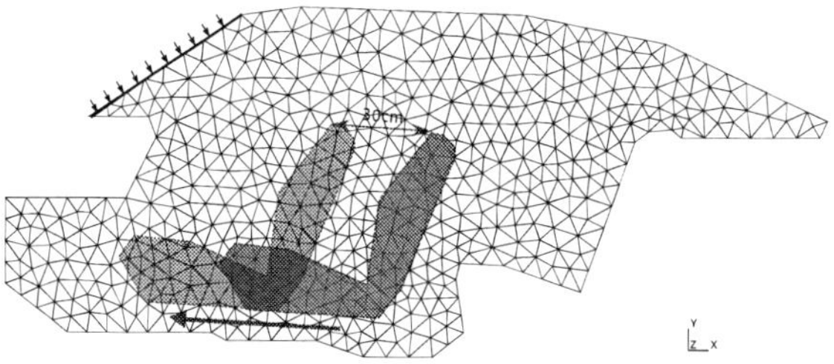

Figure 1: X-FEM mesh with level-set defined seat for stated problem

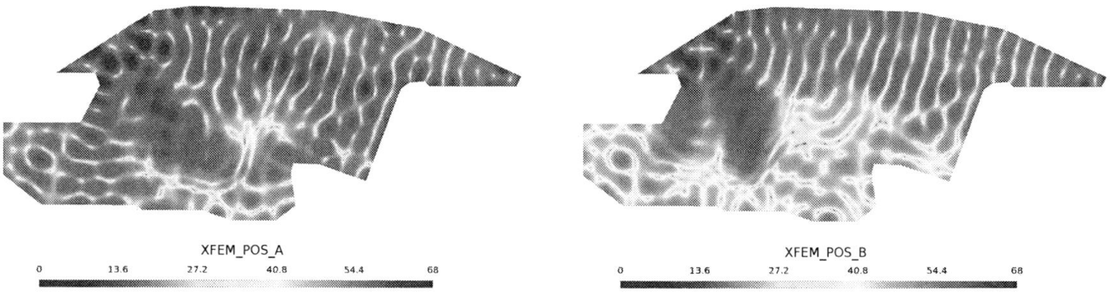

(a) (b)

Figure 2: Solution of the sound pressure (in dB) for position blue (left) and red (right)

example demonstrates the applicability and the computational convenience of the proposed method in a practical application.
Such a complex problem is solved using a relatively simple mesh. We believe that our method could be a competitive computational tool for analysing poro-elastic material involving complex geometries, even for the topological design of absorbing materials. More potential of the method is explored in the authors' current work such as "imperfect" medium interfaces behaving as strong discontinuities.

References

[1] Noureddine Atalla, Raymond Panneton, and Patricia Debergue. A mixed displacement-pressure formulation for poroelastic materials. *J. Acoust. Soc. Am.*, 104(3):1444–1452, 1998.

[2] Nicolas Dauchez, Sohbi Sahraoui, and Noureddine Atalla. Convergence of poroelastic finite elements based on Biot displacement formulation. *J. Acoust. Soc. Am.*, 109(1):33–40, 2001.

[3] G. Gabard and O. Dazel. A discontinuous Galerkin method with plane waves for sound-absorbing materials: plane-wave DGM for poro-elastic material. *Int. J. Numer. Meth. Engng*, 104(12):1115–1138, 2015.

[4] Nicolas Moes, John Dolbow, and Ted Belytschko. A finite element method for crack growth without remeshing. *Int. J. Numer. Meth. Engng*, 46:131–150, 1999.

[5] N. Moës, M. Cloirec, P. Cartraud, and J.-F. Remacle. A computational approach to handle complex microstructure geometries. *Comput. Methods Appl. Mech. Engrg.*, 192(28-30):3163–3177, 2003.

Adaptive high order Finite Element modelling of poroelastic materials

S. Jonckheere[1,4], H. Bériot[1], O. Dazel[2], W. Desmet[3,4]

[1] Siemens Industry Software, Simulation and Test Solutions,
Interleuvenlaan 68, B-3001 Heverlee, Belgium
[2] Le Mans Université, LAUM UMR CNRS 6613,
Le Mans, 720000, France
[3] Flanders Make, Core Lab DMMS
[4] KU Leuven, Department of Mechanical Engineering,
Celestijnenlaan 300, B-3001, Heverlee, Belgium

This paper presents a physics-based adaptivity algorithm to determine the directional element orders for poroelastic material models built-up through the ($\mathbf{u}^s$,p^f)*-formulation of the Biot equations. The methodology is applied to a use case to determine the plane wave absorption coefficient of a poroelastic meta-material with an elastic inclusion.*

1 Introduction

With ever more stringent requirements on noise and vibration levels, engineers strongly rely on passive treatments (e.g. multi-layered stacks of of elastic, viscoelastic and poroelastic layers) to improve the sound and vibration properties of a product. Over the past decades, CAE tools built upon the Finite Element Method (FEM) have become indispensable tools to predict the behaviour of virtual prototypes in an early design stage. Through its innate element discretisation concept, the FEM can tackle geometrically complex problem cases. Within these elements, polynomial approximation functions are used to describe the dynamic field variables. This procedure is a strength of the FEM, but it constitutes also a limitation. As the frequency increases, the number of low-order elements required to control the interpolation and pollution errors increases more than linearly. In order to balance accuracy and computational cost, frequency-dependent input meshes are often required. On top of this, poroelastic materials pose additional constraints as they have a high number of DOFs per node and support relatively short wavelengths and outspoken near-field damping effects.
To overcome these problems, research been done over the past years to improve the efficiency of FE methods. Initial steps into p-refinement (order refinement) were made by Hörlin [3] and Rigobert [6]. However, due to the innate coupling between both phases, together with the highly dispersive properties of the different wave types that propagate in a poroelastic material, it still is still highly challenging to derive simple guidelines to create a valid mesh (and FE model). Hereto, this paper proposes a higher order, automated adaptive FE procedure for poroelastic material modeling. The element orders are determined *a priori*, adapted to the frequency and based on the physics of poroelastic materials. Furthermore, these element orders can be assigned independently in different directions, in order to account for anisotropy in the mesh. The result is a (nearly) mesh-independent solution that allows to reach a user-defined target error while minimizing the computational cost.

2 Adaptive high order finite element model

The variational formulation of the Biot equations is discretised using the high-order finite element method [2, 4]. On each element, the numerical solution is described using Lobatto shape functions. These form a 'hierarchic' basis. It is therefore simple to handle models with different polynomial orders in the element. Furthermore, the hierarchic p-FEM scheme allows to enforce directional orders in the interpolation basis.
The directional element orders are built up bottom-up from the edge orders using specific element conformity rules [1] to determine the 2D- and 3D elements' directional orders. The element edge orders are determined by comparing the accuracy at given order with the solution of a reference problem, describing the propagation at oblique incidence at an angle θ, which is the most general possible problem on a 1D element topology. The dependence in the element direction to the parameter θ is low, so a value of 30° is used. The element edge orders p_{solid} and p_{fluid} are chosen to minimise the computational cost, subject to an error treshold E_T. Since the computational cost is directly influenced by the element order, the optimisation problem can be translated to minimising the weighted sum of solid- and fluid phase orders. However, the solution of this optimisation problem is not trivial. Nevertheless, in a first step the monotonous convergence of the FEM with increasing order can be

exploited. Therefore p_{solid} and p_{fluid} are increased simultaneously until $E_{L^2} < E_T$. This accords with the method of steepest descent. Thereafter, either p_{solid} or p_{fluid} can be reduced by tracing the error threshold isoline $E_{L^2} = E_T$ horizontally or vertically. p_{solid} is reduced first (if possible) as it has the highest influence on the computational cost. This strategy ensures that the set E_T is met and the computational cost is minimised.

3 Use Case: Poroelastic meta-material with elastic inclusion

The acoustic plane wave absorption in the normal direction of a poroelastic metamaterial with an elastic inclusion [7] is studied as an application. Figure 1(a) shows the predicted acoustic absorption curves for three different meshes. For all, the results are in excellent agreement with the reference model and globally satisfy the set tolerance $E_T = 1\%$ as the error curves in Figure 1(b) indicate. Figure 1(c) shows a deeper insight in the adaptive order structure for the model at 10 kHz, showing the minimal and maximal orders for both the displacement and pressure variables.

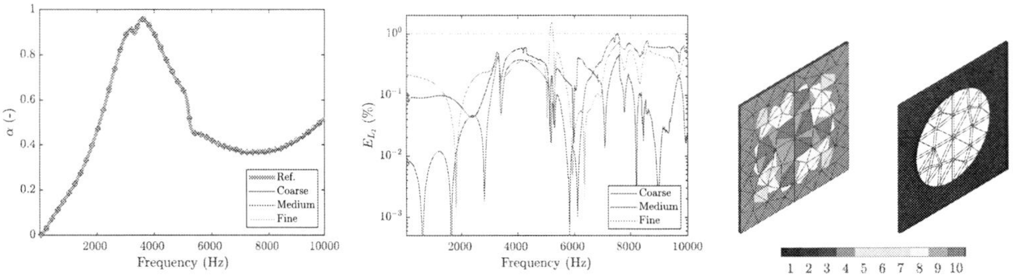

(a) Absorption coefficient (b) L^2-error w.r.t reference (c) **u** and p orders

Figure 1: Results for the absorption analysis of a poroelastic meta-material with inclusion.

4 Conclusion

This paper focuses on an efficient implementation of a higher-order Finite Element Method for poroelastic materials described by the Biot theory. Crucial is a criterion based on a 1.5D reference problem to determine the solid- and fluid phase orders semi-independently, minimising the computational cost for a required accuracy. The calculated edge orders can are then propagated to the face (2D) and solid (3D) directional element orders by using some element-type specific conformity rules. The approach is applied with success to the absorption analysis of a poroelastic meta-material with inclusion.

Acknowledgements

The research of S. Jonckheere is funded by a grant of VLAIO, the Flanders Innovation & Entrepreneurship Agency. The Research Fund KU Leuven is gratefully acknowledged for its support.

References

[1] H. Bériot and G. Gabard. Anisotropic adaptivity of the p-fem for time-harmonic acoustic wave propagation. *J. Comput. Phys.*, 378:234–256, 2019.

[2] H. Bériot, A. Prinn, and G. Gabard. Efficient implementation of high-order finite elements for Helmholtz problems. *Int. J. Numer. Methods Eng.*, 106(3):213–240, 2016.

[3] N.-E. Hörlin. 3D hierarchical *hp*-FEM applied to elasto-acoustic modelling of layered porous media. *J. Sound Vib.*, 285(1-2):341–363, 2005.

[4] S. Jonckheere, H. Beriot, O. Dazel, and W. Desmet. Efficient adaptive order poroelastic material modelling within modal vibro-acoustic system models. In *Proceedings of the Conference on Noise and Vibration Engineering 2020 (ISMA2020)*, Leuven, (Belgium), 2020.

[5] S. Rigobert, N. Atalla, and F.C. Sgard. Investigation of the convergence of the mixed displacement-pressure poroelastic materials using hierarchical elements. *J. Acoust. Soc. Am.*, 114:2607–2617, 2003.

[6] T. Weisser, J.-P. Groby, O. Dazel, F. Gaultier, E. Deckers, S. Futatsugi, and L. Monteiro. Acoustic behavior of a rigidly backed poroelastic layer with periodic resonant inclusions by a multiple scattering approach. *J. Acoust. Soc. Am.*, 139(2):617–629, 2016.

A tunable architectured core material exhibiting auxetic behaviour for sound transmission control

H. Mao[1,2], R. Rumpler[1,2], M. Gaborit[1,2], P. Göransson[1,2]

[1]The Marcus Wallenberg Laboratory for Sound and Vibration Research (MWL), Department of Engineering Mechanics, KTH Royal Institute of Technology, SE-100 44, Stockholm, Sweden
[2]The Centre for ECO2 Vehicle Design, SE-100 44 Stockholm, Sweden

Metamaterials, exhibiting exotic properties by artificial geometry design, have received increasing attention in the past decades. In this contribution, a tunable mechanical metamaterial exhibiting three-dimensional (3D) auxetic (negative Poisson's ratio) properties is proposed. The negative Poisson's ratio arises from the chirality of the cellular micro-structures, which also leads to anisotropic elastic material properties to be characterised by an inverse estimation method developed by the authors. The results show that the anisotropic mechanical properties and the Poisson's ratios in all three dimensions are tunable by simply adjusting the cellular geometry. These adjustable properties offer broad potential for applications in the fields of sensors, tissue engineering, or vibro-acoustics. The paper mainly illustrates an application on sound transmission loss of a sandwich panel in the low frequency range. The simulated results highlight the potential of the proposed approach to control the design of structures with exceptional noise insulation properties.

1 3D auxetic cell and material modelling

An isometric Kelvin cell (KC) arrangement exhibiting cubic material symmetry is chosen as the reference micro-structure, see Fig. 1a. The geometrical modifications applied to the reference KC geometry are in the form of rigid body rotations of the square faces, while keeping these square undeformed during the rotation [2]. Fig. 1b shows an auxetic cell generated from twisting all six square faces with an angle of 90°. Topologically, the cell is not modified and all vertices remain connected through the same struts (with altered lengths) as in the KC geometry. The rotational angles applied on each pair of square faces are anti-symmetric, *i.e.* rotated with opposite angles, equal in absolute value. The twisted cell exhibits chirality with respect to each middle plane parallel to the pairs of twisted faces, see Fig. 1b. When a load is applied to the chiral structure, the non-centered force or non-affine deformation induce auxetic behavior where the lateral squares from any chosen pair of squares contract laterally under compression of the cell applied to the chosen pair.
The anisotropic elastic material properties associated with this KC geometry with rotated faces are characterised by an inverse estimation method developed by the authors [3]. The corresponding results show that the estimated Poisson's ratio of the twisted cell vary as a function of the twisting angle and geometry size. A negative Poisson's ratio is found for a twisting angle in the range 35° – 90° when twisting all six square faces anti-symmetrically with a same angle.

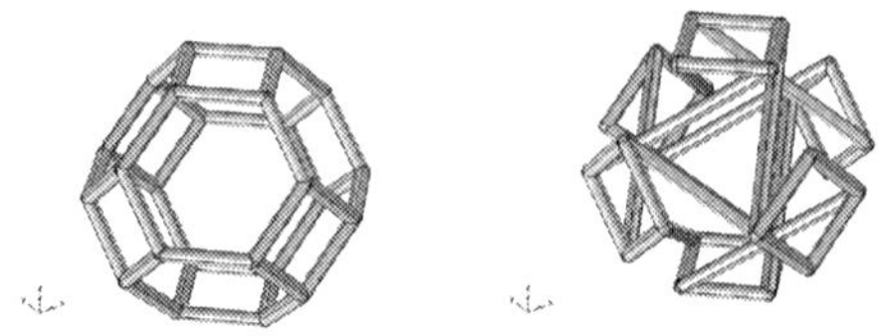

(a) (b)

Figure 1: a) Kelvin cell, b) twisted auxetic cell (90°).

2 Applications: an example in vibro-acoustics

In connection with the specific auxetic and chiral properties previously introduced, multiple applications are currently under study. Here we focus on an application in vibro-acoustics for sound insulation purposes, focusing on the sound transmission loss of a sandwich panel with a KC-based structural foam as core material. Fig. 2 shows three mode shapes of the auxetic single cell, which might be interesting for vibro-acoustics purposes. Fig. 2a shows a spring-mass-like functional mode shape which may be introduced to exhibit an internal-resonator-like behaviour, previously studied to design metamaterial sandwich structures for noise insulation purposes [1]. Fig. 2b shows an extension-torsion coupled mode shape, which may generate strong local damping effects. Fig. 2c shows the unit cell auxetic behaviour due to the chiral torsional effect under compression.

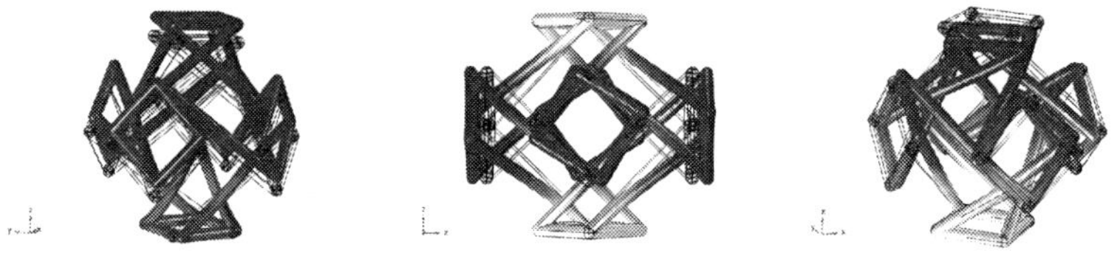

(a) (b) (c)

Figure 2: Selected interesting mode shapes of a single twisted cell: a) a spring-mass-like system, b) extension-torsion coupling in transverse direction, c) chiral torsional mode shape with auxetic effect.

Introducing such structural foams as a core material, Fig. 3a shows the finite element model of a metamaterial, periodic sandwich plate with infinite lateral dimensions. A cellular array consisting of two auxetic cells in the thickness are assembled as a meta-structure core material of the sandwich plate. Fig. 3b illustrate the resulting trend of the STL in the frequency range up to 5000 Hz. The case without internal resonator is plotted in solid line, and the potential associated with the addition of the Kelvin Cell array corresponds to the dashed line. The coincidence region is clearly visible on the reference calculation without the cellular array. A tuned cell array offers the possibility to imrove the STL in this region, at the expense of an additional deep, which itself may be mitigated by a specifically tuned resonator. Such configurations will be further demonstrated by controlling microstructure properties such as twisting angles, size and material properties of the struts.

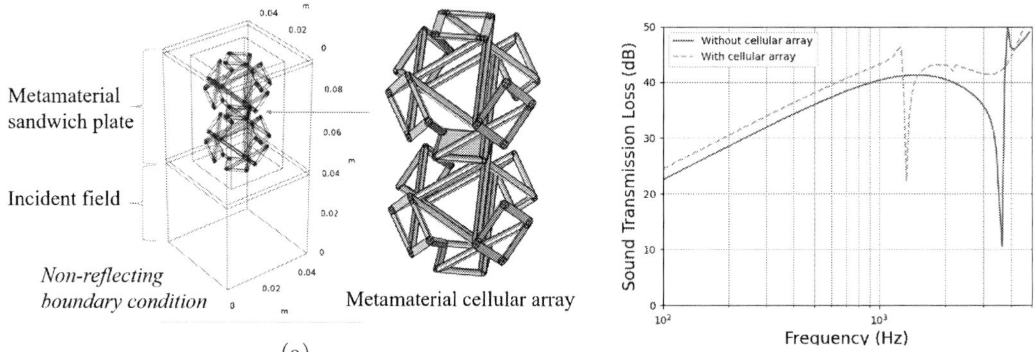

(a) (b)

Figure 3: The finite element model of the metamaterial sandwich plate, b) Sound transmission loss (STL) of the metamaterial single-layer plate.

References

[1] Zibo Liu, Romain Rumpler, and Leping Feng. Broadband locally resonant metamaterial sandwich plate for improved noise insulation in the coincidence region. *Composite structures*, 200:165–172, 2018.

[2] Huina Mao, Romain Rumpler, Mathieu Gaborit, Peter Göransson, John Kennedy, Daragh O'Connor, Daniel Trimble, and Henry Rice. Twist, tilt and stretch: From isometric kelvin cells to anisotropic cellular materials. *Materials & Design*, 193:108855, 2020.

[3] Huina Mao, Romain Rumpler, and Peter Göransson. An inverse method for characterisation of the static elastic Hooke's tensors of solid frame of anisotropic open-cell materials. *International Journal of Engineering Science*, 147:103198, 2020.

SAPEM 2020+1
6TH TRIENNIAL
SYMPOSIUM ON THE ACOUSTICS OF
PORO-ELASTIC MATERIALS

Effects of discontinuous impedance boundaries on sound absorption coefficient measurements in a standing wave tube

Huading Lou[1,2], Hequn Min[1,2*]
1. School of Architecture, Southeast University, Nanjing, 210096, China;
2. Key Laboratory of Urban and Architectural Heritage Conservation of Ministry of Education, Southeast University, Nanjing, 210096, China

Abstract

The standing wave tube (SWT) method is widely used for measuring sound absorption coefficients of acoustic materials. The effects of circumferential edge constraint[1,2] or air gaps[3,4] between the sample and the SWT on sound absorption coefficient measurements were studied previously. While sample sizes are significantly smaller than that of the SWT cross section, the measured sound absorption coefficient of this discontinuous impedance boundary (DIB) may be smaller than its actual value. Based on the benchmark theory of porous material impedance proposed by K. Attenborough[5], in this paper, we investigate the influence from DIB on absorption coefficient measurements in the SWT based on the finite element method. Two types of DIB are investigated in Fig. 1. In Fig. 1(a), the sample width and length are both smaller than those of the SWT cross section ($0.5W < a < W$, $0.5W < b < W$). While in Fig. 1(b), there is one side of the sample equal to that of the SWT ($a=W$, $0.5W < b < W$). Results in Fig. 2(e) show that measured sound absorption coefficients are sensitive to positions of the sample at medium and high frequencies, and absorption attenuation at high frequencies decreases when the sample is placed at the center of the SWT. It is also shown that different sample shapes with a same area (see (a) and (d) in Fig. 2) also affect the measured sound absorption coefficients at high frequencies. This paper can be the basis of the measurements of actual sound absorption coefficient on the condition of sample sizes significantly smaller than that of the SWT cross section.

* Corresponding author.
E-mail address: hqmin@seu.edu.cn

Key words: Discontinuity impedance boundaries; Sound absorption; Standing wave tube

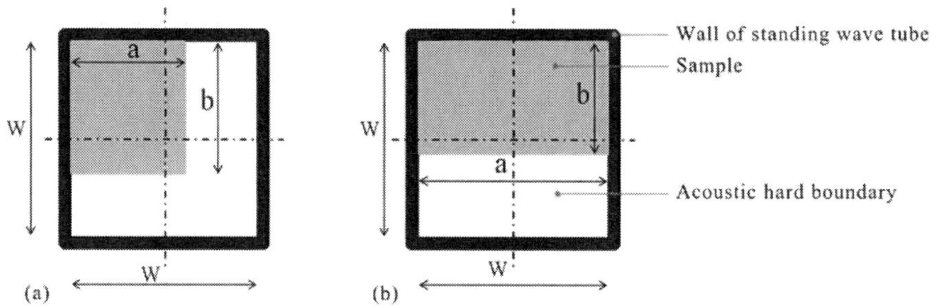

Figure 1: Schematic diagrams of the relation of cross section size between porous material sample and the square SWT

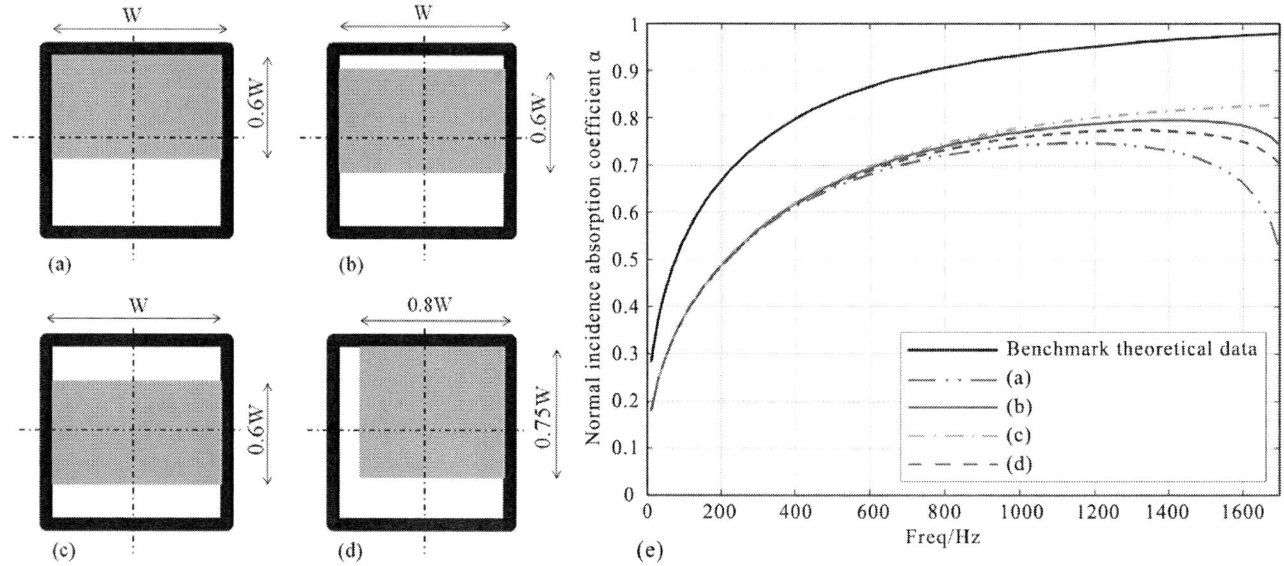

Figure 2: Effects of sample positions ((a), (b), (c)) in the SWT and shape (d) on sound absorption coefficient (e) of DIB. (Note: the areas of (a), (b), (c), and (d) are equal)

References

[1] B. H. Song and J. S. Bolton, Effect of circumferential edge constraint on the acoustical properties of glass fiber materials, J. Acoust. Soc. Am. **110**(6), pp. 2902-2916, 2001.

[2] D. Pilon and R. Panneton, Behavioral criterion quantifying the edge-constrained effects on foams in the standing wave tube, J. Acoust. Soc. Am. **114**(4), pp. 1980-1987, 2003.

[3] A. Cummings, Impedance tube measurements on porous media: the effects of air-gaps around the sample, J. Sound Vib. **151**(1), pp. 63-75, 1991.

[4] D. Pilon, R. Panneton, and F. Sgard, Behavioral criterion quantifying the effects of circumferential air gaps on porous materials in the standing wave tube, J. Acoust. Soc. Am. **116**(1), pp. 344-356, 2004.

[5] K. Attenborough. Ground parameter information for propagation modeling, J. Acoust. Soc. Am. **92**, pp. 418–427, 1992.

A non-locally reacting Impedance Model for Periodic Noise Control Treatments

A. Parrinello[1], L. Alimonti[1], N. Atalla[2]

[1] ESI North America, 12555 High Bluff Dr., Suite 160
San Diego, CA 92130, USA
[2] Université de Sherbrooke, 2500 Boulevard de l'Université
Sherbrooke, QC J1K 2R1, Canada

A Transfer Matrix (TM) based Noise Control Treatments (NCTs) modeling is proposed. Assuming each layer of the NCT to be periodic, a Generalized TM Method (GTMM) based on a Bloch modes basis reduction of the Finite Element (FE) model of the Unit Cell (UC) of each layer is employed to model the NCT in the wavenumber space and a inverse Discrete Fourier Transform (DFT) is exploited to build the impedance at the interface of the treatment with a fluid or a structure.

1 Introduction

Classical modeling techniques for NCTs are either based on detailed FE methodologies or simplified approaches. These last mainly rely on locally reacting impedance models obtained from testing or fast calculations performed by means of transfer matrix based analytical formulations assuming flat and homogeneous layers [1]. Although very efficient, this class of approaches may lack of accuracy for complex NCTs. To overcome this limitation, the in-plane heterogeneity of the treatment is taken into account by means of a GTMM [2] and the non-local behavior at discrete locations on the interface of the NCT is captured by means of an inverse DFT. The GTMM assumes each layer of the NCT to be periodic and, therefore, fully characterized by a UC. The latter is modeled by standard FE and the passage of Bloch waves, including the effect of the curvature, is imposed as boundary conditions. As a result, a TM can be obtained for each layer and then assembled to construct the TM of the NCT.

2 Theory

A NCT lying on the xy-plane has thickness h along the z-axis direction and is connected to a semi-infinite media (*e.g.* a fluid) or another system (*e.g.* a structure) through a rectangular surface of size $a \times b$ located at $z = h$. We are interested in the dynamic stiffness matrix, $\mathbf{D}$, of the NCT at the interface ($z = h$) when generic boundary conditions (*e.g.* semi-infinite fluid, hard wall, etc ...) are applied at the opposite side ($z = 0$). Assuming the NCT to be (i) periodic (of infinite extent) in the xy-plane, (ii) defined by its UC and (iii) subjected to an in-plane periodic motion with periods a and b, an inverse DFT can be used to map the NCT stiffness evaluated for a set of n discrete wavenumbers in the xy-plane on a lattice of n discrete locations over the interface area ($z = h$). In case of an unwrapped curved NCT, the wavenumber components, k_x and k_y, are defined on the curvilinear coordinates and the curvature is accounted for in the UC dynamics. In order to obtain a proper DFT operator, the wavenumber sampling vectors must have the following form

$$\mathbf{k}_x = \frac{2\pi}{a}\mathbf{O}_{\hat{M}_y} \otimes \left[-M_x \ \dots \ M_x \right] \ , \ \mathbf{k}_y = \frac{2\pi}{b}\left[-M_y \ \dots \ M_y \right] \otimes \mathbf{O}_{\hat{M}_x} \tag{1}$$

where $\otimes$ denotes the Kronecker product, $\mathbf{O}_n$ is a row vector of ones of size n, M_x and M_y are integers defining the maximum orders of wavenumbers considered, $\hat{M}_x = 2M_x + 1$ and $\hat{M}_y = 2M_y + 1$. The lattice has spacings $l_x = a/\hat{M}_x$ and $l_y = b/\hat{M}_y$. In case the dynamics of the NCT need to be projected on the Bloch modes, the interface must embrace an odd number of UCs and the wavenumber vectors must be enriched with the Bloch modes as

$$\mathscr{X} = \mathbf{k}_x \otimes \mathbf{O}_N + \mathbf{O}_M \otimes \mathbf{n}_x \ , \ \Upsilon = \mathbf{k}_y \otimes \mathbf{O}_N + \mathbf{O}_M \otimes \mathbf{n}_y \ , \tag{2}$$

where $M = \hat{M}_x\hat{M}_y$ is the total number of cells, $N = \hat{N}_x\hat{N}_y$ is the total number of Bloch modes, $\hat{N}_x = 2N_x + 1$, $\hat{N}_y = 2N_y + 1$, N_x and N_y are the maximum orders of Bloch modes selected and the Bloch modes vectors are defined as

$$\mathbf{n}_x = \frac{2\pi}{l_x}\mathbf{O}_{\hat{N}_y} \otimes \left[-N_x \ \dots \ N_x \right] \ , \ \mathbf{n}_y = \frac{2\pi}{l_y}\left[-N_y \ \dots \ N_y \right] \otimes \mathbf{O}_{\hat{N}_x} \ . \tag{3}$$

The corresponding lattice coordinates are defined as $\mathbf{x} = \mathscr{X} a^2/(2\pi \hat{M}_x \hat{N}_x)$ and $\mathbf{y} = \Upsilon b^2/(2\pi \hat{M}_y \hat{N}_y)$. The dynamic problem is first solved in the wavenumbers space as $\hat{\mathbf{D}}\hat{\mathbf{q}} = \hat{\mathbf{F}}$, where $\hat{\mathbf{q}}$ and $\hat{\mathbf{F}}$ are the vectors of generalized displacement and forces collecting Bloch modes components and $\hat{\mathbf{D}}$ is a block diagonal matrix whose i-th block, $\hat{\mathbf{D}}^{(i)}$, is the stiffness matrix of size Nm related to the i-th couple of wavenumbers from vectors $\mathbf{k}_x$ and $\mathbf{k}_y$ and projected on the related Bloch modes and m is the number of degrees of freedom for each lattice node (1 for fluids, 3 for solids, 4 for porous). In case of a NCT defined as a generic arrangement of periodic layers, the TMs of all layers, $\hat{\mathbf{T}}_j^{(i)}$, must be first defined and the global TM, $\hat{\mathbf{T}}^{(i)}$, must be assembled as described in [2]. Then, the problem must be recast in stiffness form, thus obtaining $\hat{\mathbf{D}}^{(i)}$. The vectors of generalized displacements and forces at the lattice points can be expressed as $\mathbf{q} = \hat{\mathbf{S}}\hat{\mathbf{q}}$ and $\mathbf{F} = \hat{\mathbf{S}}\hat{\mathbf{F}}$ respectively. The transform map is defined as $\hat{\mathbf{S}} = \exp\left(-\mathrm{j}\mathbf{x}^T \mathscr{X} - \mathrm{j}\mathbf{y}^T \Upsilon\right) \otimes \mathbf{I}_m/\sqrt{NM}$, where $\exp(*)$ denotes the element wise exponential operator and $\mathbf{I}_n$ is the identity matrix of size n. Then, the stiffness matrix at the lattice points can be expressed as $\mathbf{D} = \hat{\mathbf{S}}\hat{\mathbf{D}}\hat{\mathbf{S}}^H$.

3 Application

The sound absorption of a 20 mm thick NCT made of foam with infinite cylindrical void inclusions and backed by a rigid wall (at $z = 0$) is considered. Inclusions are parallel to the y-axis, have diameter of 15 mm and are 20 mm spaced along the x-axis. The foam has properties: $\Phi = 0.95$, $\sigma = 8900$ Nsm^{-4}, $\alpha_\infty = 1.42$, $\Lambda = 180$ μm, $\Lambda' = 360$ μm, $E = 224$ kPa, $\nu = 0.4$, $\mu = 17$ %, $\rho = 6.1$ kgm^{-3}. It is modeled as an equivalent fluid according to the Johnson–Champoux–Allard rigid model [1]. The top fluid ($z > h$) has properties $c = 340$ ms^{-1} and $\rho = 1.284$ kgm^{-3}. The UC is modeled with the Tetra4 mesh depicted in Figure 1a. The acoustic impedance at the interface with the semi-infinite fluid is evaluated at the lattice points by assuming an infinite rigid baffle. Figure 1b presents the comparison of the dissipation coefficient at normal incidence predicted by the proposed approach with the result produced by a double porosity analytical model (Reference) [3]. It can be observed that Bloch modes are essential to capture the physics of the problem, especially above 17 kHz.

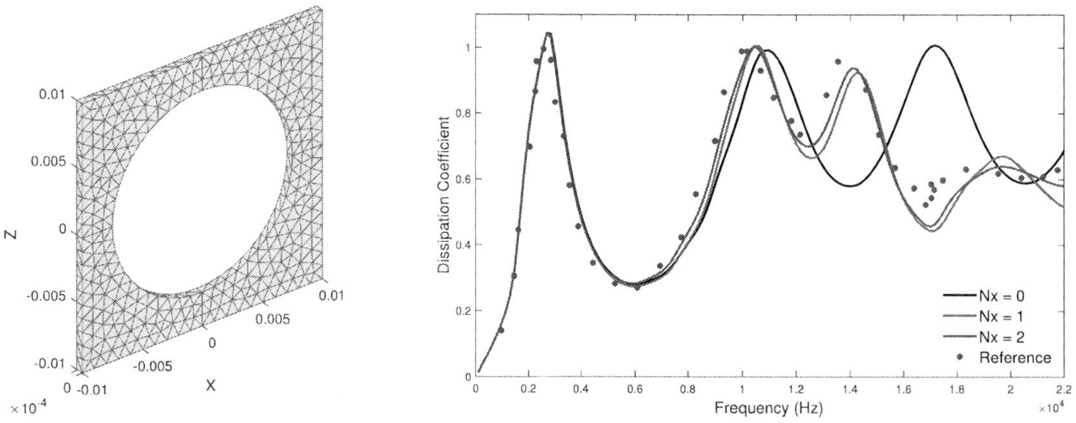

(a) Tetra4 mesh of the unit cell

(b) Dissipation coefficient at normal incidence

Figure 1: 20 mm thick NCT made of foam with cylindrical inclusions and backed by a rigid wall with in-plane dimension of 0.22 m × 0.22 m ($M_x = M_y = 10$, $N_y = 0$).

References

[1] J. Allard and N. Atalla, *Propagation of sound in porous media : modelling sound absorbing materials 2nd edition*, John Wiley & Sons, 2009.

[2] A. Parrinello, G.L. Ghiringhelli and N. Atalla, *Generalized Transfer Matrix Method for periodic planar media*, J. Sound Vib, **464**, 2020.

[3] J.-P. Groby, O. Dazel, A. Duclos, L. Boeckx, L. Kelders, *Enhancing the absorption coefficient of a backed rigid frame porous layer by embedding circular periodic inclusions*, J. Acoust. Soc. Am. **130 (6)**, pp. 3771-3780, 2011.

Purdue University and Matelys – Research Lab
585 Purdue Mall
West Lafayette, Indiana, USA
47907-2088

ISBN 978-1-7138-3084-9

9 781713 830849

CURRAN ASSOCIATES INC.
proceedings.com